Volume 28

Advances in
Librarianship

Editorial Advisory Board

Nancy H. Allen, Denver University
Danuta Nitecki, Yale University
Mary Jean Pavelsek, New York University
Nancy Roderer, Johns Hopkins University
Robert A. Seal, Texas Christian University
Jim Welbourne, New Haven Free Public Library

Volume 28

Advances in
Librarianship

81691

Edited by
Danuta A. Nitecki

Yale University Library
New Haven, Connecticut

CONCORDIA COLLEGE LIBRARY
BRONXVILLE, NEW YORK 10708

2004

ELSEVIER
ACADEMIC
PRESS

Amsterdam Boston Heidelberg London New York Oxford Paris
San Diego San Francisco Singapore Sydney Tokyo

ELSEVIER B.V.	ELSEVIER Inc.	ELSEVIER Ltd	ELSEVIER Ltd
Radarweg 29	525 B Street, Suite 1900	The Boulevard, Langford Lane	84 Theobalds Road
P.O. Box 211, 1000 AE	San Diego, CA 92101-4495	Kidlington, Oxford OX5 1GB	London WC1X 8RR
Amsterdam, The Netherlands	USA	UK	UK

© 2004 Elsevier Inc. All rights reserved.

This work is protected under copyright by Elsevier Inc., and the following terms and conditions apply to its use:

Photocopying
Single photocopies of single chapters may be made for personal use as allowed by national copyright laws. Permission of the Publisher and payment of a fee is required for all other photocopying, including multiple or systematic copying, copying for advertising or promotional purposes, resale, and all forms of document delivery. Special rates are available for educational institutions that wish to make photocopies for non-profit educational classroom use.

Permissions may be sought directly from Elsevier's Rights Department in Oxford, UK: phone (+44) 1865 843830, fax (+44) 1865 853333, e-mail: permissions@elsevier.com. Requests may also be completed on-line via the Elsevier homepage (http://www.elsevier.com/locate/permissions).

In the USA, users may clear permissions and make payments through the Copyright Clearance Center, Inc., 222 Rosewood Drive, Danvers, MA 01923, USA; phone: (+1) (978) 7508400, fax: (+1) (978) 7504744, and in the UK through the Copyright Licensing Agency Rapid Clearance Service (CLARCS), 90 Tottenham Court Road, London W1P 0LP, UK; phone: (+44) 20 7631 5555; fax: (+44) 20 7631 5500. Other countries may have a local reprographic rights agency for payments.

Derivative Works
Tables of contents may be reproduced for internal circulation, but permission of the Publisher is required for external resale or distribution of such material. Permission of the Publisher is required for all other derivative works, including compilations and translations.

Electronic Storage or Usage
Permission of the Publisher is required to store or use electronically any material contained in this work, including any chapter or part of a chapter.

Except as outlined above, no part of this work may be reproduced, stored in a retrieval system or transmitted in any form or by any means, electronic, mechanical, photocopying, recording or otherwise, without prior written permission of the Publisher.

Address permissions requests to: Elsevier's Rights Department, at the fax and e-mail addresses noted above.

Notice
No responsibility is assumed by the Publisher for any injury and/or damage to persons or property as a matter of products liability, negligence or otherwise, or from any use or operation of any methods, products, instructions or ideas contained in the material herein. Because of rapid advances in the medical sciences, in particular, independent verification of diagnoses and drug dosages should be made.

First edition 2004

ISBN: 0-12-024628-7
ISSN: 0065-2830 (Series)

♾ The paper used in this publication meets the requirements of ANSI/NISO Z39.48-1992 (Permanence of Paper). Printed in Great Britain.

Working together to grow
libraries in developing countries

www.elsevier.com | www.bookaid.org | www.sabre.org

ELSEVIER   BOOK AID International   Sabre Foundation

Contents

Contributors ix
Preface xi

Evolving Spaces: An Architect's Perspective on Libraries   1
Alexander P. Lamis

    I. Introduction   1
   II. Centrifugal Force   2
  III. The Development of Modern Libraries   3
  IV. Libraries and the City Beautiful Movement in America   5
   V. Modular Design in Libraries   6
  VI. The Post-modern Reaction   7
 VII. Today's Library Design Issues   8
VIII. Conclusion   14
       References   15

Excellent Libraries: A Quality Assurance Perspective   17
Felicity McGregor

    I. Introduction   17
   II. The Search for Excellence   19
  III. Defining Quality   21
  IV. Adopting a Business Excellence Framework   22
   V. Libraries as Business Organisations   26
  VI. Challenges and Insights   33
 VII. Organisational Culture   38
VIII. Benchmarking   42
  IX. Success and Sustainability   45
   X. Conclusion   51
       References   52

The Civic Library: A Model for 21st Century Participation   55
Diantha Schull

    I. Introduction   55
    II. Context   56
    III. Paradox   59
    IV. Need for a Service Model   60
    V. The LFF Civic Library Model   63
    VI. Civic Libraries in Practice   68
    VII. Conclusion   80
        References   80

Libraries and Learning   83
Robert S. Martin

        References   93

The Evolving Relationships between Libraries and Scholarly Publishers: Metrics and Models   95
Craig Van Dyck and Christopher McKenzie

    I. Introduction   95
    II. How Publishers Decide What to Publish   96
    III. How Publishers Market and Sell Content   101
    IV. Usage Data as a Metric   109
    V. Areas of Potential Cooperation and Collaboration   112
    VI. Conclusions   118
        Acknowledgements   118
        References   119

The US Government and E-Government: Two Steps Forward, One Step Backwards?   121
Peter Hernon and Robert E. Dugan

    I. Overview   122
    II. Issues   134
    III. Definite Barriers to Information Access
        (One Step Backwards)   139

IV. A Modest Research Agenda   140
V. Implications of E-Government to Libraries   143
VI. Conclusion   145
  References   147

Information Seekers' Perspectives of Libraries and Librarians   151
Eileen Abels

I. Introduction   151
II. Background   152
III. Information Seeking Behaviors Today   156
IV. MBA Students' Perceptions of Librarians and Libraries: A Case Study   161
V. Discussion   164
VI. Conclusions   167
  References   168

Competition or Convergence? Library and Information Science Education at a Critical Crossroad   171
Joan C. Durrance

I The Changed Information Landscape   171
II. It is Not Just Technology: The Changing Research Paradigm   172
III. Operating in a Highly Competitive Environment   173
IV. Documenting LIS Education in the Midst of Change   174
V. The KALIPER Project   175
VI. KALIPER-Identified Trends and Their Current Manifestation   177
VII. Information Education: Competition or Convergence?   191
  References   196

Index   199

Contributors

Numbers in parantheses indicate the pages on which the authors' contributions begin.

Eileen Abels (151), College of Information Studies, University of Maryland, College Park, MD 20742, USA

Robert E. Dugan (121), Sawyer Library, Suffolk University, 8 Ashburton Place, Boston, MA 02108, USA

Joan C. Durrance (171), Margaret Mann Collegiate Professor of Information, School of Information, University of Michigan, Ann Arbor, MI, USA

Peter Hernon (121), Graduate School of Library and Information Science, Simmons College, 300 The Fenway, Boston, MA 02115-5898, USA

Alexander P. Lamis (1), Robert A.M. Stern Architects, New York, NY, USA

Robert S. Martin (83), Institute of Museum and Library Services, Washington, DC, USA

Felicity McGregor (17), University of Wollongong Library, Wollongong, NSW, Australia

Christopher McKenzie (95), Scientific, Technical, and Medical Publishing, John Wiley & Sons, Inc., Hoboken, NJ, USA

Diantha Schull (55), Americans for Libraries Council, New York, NY, USA

Craig Van Dyck (95), Scientific, Technical, and Medical Publishing, John Wiley & Sons, Inc., Hoboken, NJ, USA

Preface

Different Perspectives on Libraries and Librarianship

Advances in Librarianship is a series with a long-standing reputation as a resource for current reviews of major issues or themes facing librarians and the institutions in which they practice their profession. The series is characterized by its collection of in-depth chapters offering broad coverage of the library field over time, and an updated snapshot of specific, sometimes recurring, themes in a given volume. In recent years, several volumes acknowledge the influence of automation on libraries, while not ignoring developments in the core topics of concern to librarians such as preservation, cataloging, collections and services.

With preparing this 28th volume, I joined the ranks of the series' editors. I vowed to continue the series' traditions of attracting distinguished authors in librarianship to prepare original writings and of presenting in one volume a collection of thoughtful reflections and useful reviews of subjects that will be stimulating to those concerned about the life of the library, its mission, and the challenges facing its professional stewards. In this inaugural volume, however, I proposed to break slightly from tradition and instead of seeking library colleagues to document what the profession is doing, I sought to capture different perspectives on the library and librarianship, particularly to record how others view the profession and contributions of libraries to society and culture.

Taking an external view of libraries and librarianship has its own challenges, not least of which is to recruit writers to share a perspective from outside the profession or beyond focusing on operational details. Members of an excellent Editorial Board joined me in identifying and approaching experts both who share a stake in the future of the library and a keen interest in what librarianship offers, and who took a view from outside the field, and by doing so, have provided us with a new set of ideas to advance librarianship.

We posed a number of questions to prospective authors in our invitations to contribute to this volume. We asked an architect how he viewed the library as a physical place, and what challenges him in designing space to house collections, integrate technologies, and create environments for reading, reflection, and learning. We wondered what insights a library director and juror of a national quality award had about the library as a business organization striving for excellence. We asked an executive director of a non-profit advocacy group of trustees about the library's civic role, and a director of federal funding about the library's collaborations with other cultural institutions. We asked publishers how electronic publishing and pricing models have changed their view of libraries as customers. We turned to library researchers to ask what relationship the library has to those who use it—are libraries seen as a primary source for both information and assistance in finding it? We asked a library educator for perspectives on the seeming competition between library and information science in preparing future professionals to carry the value librarianship has to individuals, society, and other fields engaged in the communication of information over time and across geography.

The result of our inquiries is a collection of eight chapters, each uniquely addressing different perspectives on libraries and librarianship. There was no attempt to coordinate the perspectives and each author knew little or nothing about what the others were addressing. Yet, there are a few insights repeated among the chapters. Abels, Van Dyck, and McKenzie confirm the premise of this volume when they separately ask rhetorical questions in their two chapters to which they almost identically respond that the answer "depends on the perspective of who is asking and who is answering." Most of the authors cite the tremendous impact of information technologies and the Internet on libraries, while acknowledging that librarianship has not lost sight of its role to preserve powerful cultural and societal values of assuring that people can access information for their endeavors, for life-long learning and ultimately for improving the quality of life.

The arrangement of these chapters is less important than our invitation to read them all, in whatever sequence. They are presented here loosely to reflect three broad perspectives—the library viewed as an organization occupying a physical place and holding a civic function, the library experienced as a partner with other institutions engaged in scholarly and cultural communications, and the library as a choice for assistance by those seeking information. The final chapter provides a perspective on librarianship as a goal of professional education.

The *library as place* frequently is cited as a cherished metaphor, regardless of the type of library or the age or occupation of those served by it. With an interest in the physical and cultural context within which library buildings are designed and built, Alexander P. Lamis, partner in the celebrated New York architectural firm of Robert A. M. Stern and Associates, reviews in Chapter 1, the history of library building development, noting the influence of various changes such as information and communication technologies and ways to shelve books. Woven in this historic review of library design issues are the challenges of libraries as expressions of social context. Mr. Lamis views the library, along with information centers and classroom buildings, as "structures that house the apparatus of the Information Age." He concludes his historic review with reflections on contemporary expectations for library buildings to provide an environment sought for storage of materials and use by readers.

The library is also a place where people work and thus is viewed by some as a *business organization*. In Australia, the principles of quality improvement and assurance are fast gaining inclusion as part of the work and service cultures, with national quality award recognition bestowed through competitive review. The first library in Australia, and perhaps worldwide, that has successfully competed with a variety of profit and not-for-profit organizations for a national Business Excellence Award is the University of Wollongong Library. Its director, Felicity McGregor, shares in Chapter 2, insights into the application of a business management model to libraries and the exciting results. She reports on the library's "quality journey" in which she led staff to examine all elements of its structure, systems, services, processes and people, and transformed the inward focused academic library to be an exemplar client-oriented service organization, with data and information systems, process improvement, and emphasis on business results. The framework Ms. McGregor describes in her chapter is a holistic leadership and management system for achieving excellence.

The library, especially the public library, is viewed to be a vital *institution of a democratic society*. In Chapter 3, Diantha Schull, Executive Director of the Americans for Libraries Council, a national non-profit advocacy organization based in New York, evolves a passionate argument for a civically oriented service model that helps build communities and engages its citizenry. She takes a practical perspective on the challenges to implement the concept in practice, describing numerous examples across the country of successful library initiatives that are creating the "Civic Library" through such offerings as public space, community information, public dialogue, and public memory.

The role of the library in society is also a focus offered in Chapter 4 by Robert S. Martin, Director of the Institute of Museum and Library Services in Washington, DC. Observing that the lines that distinguish cultural heritage institutions are blurring, he proposes that the professions represented traditionally by librarians, museum curators, archivists and broadcasters might merge to collaboratively collect and organize information and foster learning environments.

Viewed from the commercial publisher's perspective, opportunities for collaborations and partnerships with libraries are advocated in Chapter 5 by Craig Van Dyck, Vice President of Operations and Christopher McKenzie,

Director of Sales for North and South America, both in Scientific, Technical, and Medical Publishing at John Wiley & Sons, Inc. in Hoboken, New Jersey. The authors examine several models and metrics that have significantly affected publishing over the past decade, with a common influence seen in the dynamic and expanding Web. They explore the role of the library in publishing decisions; changes in how information is marketed and sold to libraries in the electronic environment; the influence of new usage data on perceptions of content by both librarians and publishers; and areas of mutual interest such as standards, pricing, and preservation of electronic resources that promise shared benefits from partnering in collaborative explorations.

The relationship of the library to those who use it is explored from two different perspectives in the next two chapters. In Chapter 6, Peter Hernon, Professor of Library and Information Science at Simmons College in Boston, Massachusetts and Robert E. Dugan, Library Director at Suffolk University, also in Boston, advocates use of technology by government agencies for improving government and its responsiveness to those governed. Although not fully achieved yet today, "e-government," along with additional challenges of the Web, is transforming the traditional role of libraries to assist the public not only in identifying and retrieving government information but to engage them to review it in order to participate more in shaping public policy.

In Chapter 7, Eileen Abels, Associate Professor of Library and Information Science at the University of Maryland at College Park, challenges us to shift our library-centric perspective about "library users" toward a user-centric view of the library as one among many resources that the larger population of "information seekers" tap. She challenges libraries to think how they can become an integral channel of information seeking and the librarian a preferred personal source of assistance and information. She first describes two classic models of information seeking that portray the role of the librarian and then offers research findings about influences on information seeking behavior. Professor Abels reports the results of her research about how business graduate students perceive librarians and use libraries. Although she recognizes that information seeking is not the same for everyone, she concludes the chapter with an updated model proposing the typical information seeker's perspective of libraries and librarians.

The preparation of this volume involved several more people than the number of chapters presented. I am grateful to all the authors for writing their chapters and then for responding to the editorial suggestions I sent them based on the careful reading by two readers from the Editorial Board and myself. A special salute is extended to the wonderful members of the Editorial Board, three continuing from work on previous volumes— Nancy Allen, Dean and Director of Penrose Library, Denver University; Mary Jean Pavelsek, International Business Librarian, New York University Libraries;

Robert Seal, University Librarian, Texas Christian University; and two joining the Board this year—Nancy Roderer, Director of Welsh Medical Library, Johns Hopkins University; and James Welbourne, Director of the New Haven Free Public Library. I personally appreciate their encouragement and hard work in a variety of tasks I faced as a new editor, as well as their identification of prospective authors and carefully reviewing chapter drafts. I also thank Chris Pringle, Publisher, Social Sciences Elsevier, who gently introduced me to the publishing phases of producing such a volume and was graciously patient in allowing me to figure out how to entice authors and prepare the collection of their work for production within flexible deadlines. The authors, Editorial Board members and publisher deserve shared credit for the success of this volume, and it is my pleasure to warmly recognize that tribute.

Danuta A. Nitecki
Editor
May 31, 2004

Evolving Spaces: An Architect's Perspective on Libraries

Alexander P. Lamis
Robert A.M. Stern Architects, New York, NY, USA

I. Introduction

Architecture is a social art. Buildings reflect the social and material conditions of the place and time where they are created. This is especially true of libraries, which represent our collective aspirations. For architects and clients involved in planning libraries, it is important to take a broad view of the task at hand, to understand the premises that guide design solutions, and to place libraries within a cultural context.

There are a number of questions that should be asked at the outset of any planning or building project: What sort of "information society" do we live in? How do we learn today? What is the role of public space in a highly privatized built environment? How can libraries be flexible enough to accommodate future changes we cannot anticipate? Library buildings represent significant investments by governments and academic institutions, and the answers must be robust enough to meet the challenges of a rapidly evolving society.

We live in a period of change and rapid evolution in the design of library buildings, information centers, classroom buildings, and related structures that house the apparatus of the Information Age. The search for answers to basic questions will yield library designs that will have long-term utility in the future.

This article examines the physical and cultural context within which library buildings are designed and built, especially the tendency of ubiquitous electronic information to erode traditional building types. Some highlights in the development of library buildings are reviewed, and then current issues in library design are discussed. Both public and academic libraries are considered. While there are certainly differences between types of libraries, the article tends to focus on areas of similarity, like the increasing importance of sustainable design, rather than on differences.

II. Centrifugal Force

In high school chemistry, students use a centrifuge to separate out heavier precipitate from liquid suspension. The whirling motion of the centrifuge, which is the result of an externally applied force, causes materials to migrate away from the center of the apparatus towards its perimeter. Our built environment has similarly been stressed by centrifugal forces in the form of economic conditions; the tendency of electronic markets to exchange goods and services directly between individuals without the mediation of a physical marketplace (Negroponte, 1995; Mitchell, 1995). Integrated downtown shopping and its suburban equivalent, enclosed shopping malls, are giving way to stand-alone big-box stores, increasingly fitted out like warehouses and convenient for shopping from home. Libraries are not immune from this effect of centrifugal force, and the dematerialization of the physical and intellectual marketplace. In this environment several questions suggest themselves: How much should be invested to create buildings to store books and other materials? How much growth should be planned? Should libraries be community or student centers, or should they be more like warehouses that are focused on the efficient distribution of information? Should libraries cease to exist as physical places at all, and exist only as decentralized networks? Are there other models for the distribution of information that we should consider?

It is instructive to consider another piece of personal technology, of somewhat earlier vintage than the personal computer and the Internet, that has had a similarly destabilizing and transforming effect on the built environment: the automobile. In contrasting the dense, centralized cities along the eastern seaboard like Boston, New York, and Philadelphia, with newer cities like Houston, Phoenix, and Los Angeles, one is struck by the centrifugal influence the automobile has had on their physical form (Kunstler, 1993; Garreau, 1991).

What is even more interesting, a century after the widespread adoption of automobiles and the suburban lifestyle they enable is the degree of leveling between these two kinds of cities. In pre-automobile cities, suburban and exurban development has accelerated, with areas of exurban New Jersey and Connecticut physically indistinguishable from California. There has also been a marked, if somewhat smaller, redevelopment of the central cores of sun-belt cities as places for entertainment and culture as much as for work.[1] There is a desire on the part of many suburbanites for something that

---

[1] There are many groups investigating and promoting the redevelopment of central cities. Among the most prominent, are the Urban Land Institute (www.uli.org) and the Congress of New Urbanism (www.cnu.org).

resembles a traditional city center and the cultural institutions that help to define it. Witness the building within the past 10 years of major new downtown libraries in Jacksonville, Salt Lake City, Denver, Seattle, Nashville, and San Antonio (Ripley, 2003, pp. 96–98). These cities are seeking to reinvigorate their city centers, and see libraries as an important factor in their redevelopment. Central libraries with specialized collections and a variety of public programs are seen as a way to draw suburbanites back downtown (Albanese, 2001). This in turn can feed the development of nearby restaurants, shops, and even loft conversions, making the library not only a cultural anchor, but also a generator of economic activity. In both Nashville and Jacksonville, where Robert A.M. Stern Architects has designed the central library buildings, the library development was part of a larger investment in downtown that also included new courthouses, stadiums, arenas, symphony halls, and art museums.[2] American cities are striving to find a new balance between core and periphery.

III. The Development of Modern Libraries

Before assessing significant current design issues and trends, it is worth briefly reviewing the development of library planning and design. Library plans have changed significantly in response to broad social and economic forces: increasing democracy, increased dissemination of knowledge, advances in science and technology, and the rise of modern cities and universities.

Libraries have been repeatedly transformed by technical invention and social evolution. The advent of printed books in Europe led to explosive growth in their production in the 15th century (Lerner, 1998, pp. 96–108). It is estimated that 10 million books were printed between 1450 and 1500 (Lerner, 1998, p. 97). The physical library was permanently transformed by this development. The chains that held volumes fast in medieval cloisters disappeared, along with the art of the painstakingly illustrated manuscript. The design of book-reading rooms and whole libraries evolved through a series of bold experiments. The Bodleian Library at Oxford (built in 1610) was laid out along functional lines with books arranged along the building perimeter in rows and in alcoves (Baur-Heinhold, p. 59). At the University of Leyden (built in 1650) a large reading room contains upright book stacks arranged by subject category as in a modern reading room (Baur-Heinhold, p. 31). The reduced cost of printing led to more variety in the types of printed materials and an increase in the complexity of library forms: reading rooms dedicated to specific purposes were created as well as offices for librarians.

[2]See The Better Jacksonville Plan Website, www.betterjax.com.

Building designs were created, copied, and transformed. The Radcliffe Library at All Souls College, Oxford (built in 1747) was designed in the circular form of a rotunda, as was the Herzog August Bibliotheca in Wolfenbuttel (built in 1706) (Baur-Heinhold, p. 63 and p. 159; Thompson, 1963, p. 14). These in turn prefigured the design by Thomas Jefferson for the library on the Great Lawn at the University of Virginia, and later in the 19th century the Library of Congress, among many others. Proposals were also brought forward for idealized library designs. In Paris in 1780, the architect Etienne-Louis Boullée proposed to transform the Royal Library by creating a living embodiment of Raphael's School of Athens with an immense vaulted reading room to contain all the assembled knowledge of the world (de Ménil, 2002, pp. 62–63).

The first half of the 19th century brought about treatises on library design as a distinct building type. In 1816 an Italian, Leopoldo della Santa, proposed a prototypical library that segregated spaces for storing books, rooms for readers, and offices for staff (Leopoldo della Santa, 1816, p. 32; see also Thompson, 1963, pp. 70–71). Le Comte de Laborde (1845) published his influential *Etude de L'Organisation des Bibliothèques* in 1845 describing ideal plans for a series of library and museum buildings. In his design for a library with 200,000 volumes, book storage is separated from public reading spaces, with books efficiently laid out on a lower level and separate rooms for catalogs and journals. Readers are comfortably housed on the floor above, with books brought to them by means of stairways. Separate apartments were provided for library administrators (Planche XI, Le Comte de Laborde, 1845, p. 48).

One of the most influential building designs of the 19th century was the library that houses the books and manuscripts of the former monastery of St Genevieve in Paris by the French architect Henri Labrouste (Richards, 1977). After being awarded the Prix de Rome of the Ecole des Beaux-Arts and traveling throughout Europe, Labrouste spent a long and fruitful career designing two major library buildings in Paris, the Bibliothèque St Genevieve, and the Bibliothèque National, the national library of France, the former Royal Library that had been the object of Boullée's study half a century earlier. State patronage allowed Labrouste to work out the designs of these major cultural commissions over a long period of time and in great detail, an approach that is significantly different from current practice (Dubbini, 2002). In the Bibliothèque St Genevieve Labrouste merged modern construction materials and methods with rigorous planning and a thorough knowledge of history and precedent. Books are stored in a mezzanine level between the entry floor and a great reading room on the *piano nobile*. The reading room is formed from cast-iron columns and arched ceiling trusses, much like the great train sheds built in Paris during the same period. The use of cast-iron in combination with stone gives the reading room a sense of both solidity and lightness. The façade design, incorporating incised names of great authors whose works were found in the library, was widely admired and closely imitated 30 years later by the American architectural firm of McKim, Mead, and White in their design for the Boston Public Library on Copley Square (Moore, 1970, pp. 62–68). McKim, Mead, and White also created a great reading room, Bates Hall, spanning the length of the principal façade behind high arched windows in a similar manner to Bibliothèque St Genevieve, but without the innovative exposed iron structure.

IV. Libraries and the City Beautiful Movement in America

The Boston Public Library inaugurated a significant period of library building in America. In the period between 1890 and 1930 many American cities built significant library buildings, and many universities built central libraries as the focal point of expanding campus plans (Kaser, 1997, pp. 45–105). Large academic libraries were constructed not only at wealthy private universities, but also at land grant and other public universities; these libraries greatly exceeded in size and complexity anything that had come before. Each represented the intellectual heart of its campus and an investment in architectural design as a civic art that are without parallel in American history.

Most of the large libraries built during this period are planned in a similar way and share a common architectural expression.[3] The Baker Library at the Harvard Business School, built in 1928 to the designs of McKim, Mead, and White, may be taken as a typical example (Cruikshank, 1987, pp. 156–160). There is an imposing symmetrical façade in the Collegiate Georgian Style with a central entrance portico. The lobby is grand in scale. A flight of ascending stairs leads to a call desk on the second floor. The majority of the holdings are in a self-supporting stack core closed to the public. Materials are requested by written orders. The main reading room is provided with high sky lit ceilings and large windows oriented to maximize exposure to northern light. Readers sit at large wooden reading tables, while artificial lights, a new technology at the turn of the 20th century, augment the natural daylight. There are separate areas for newspapers, periodicals, reference books, and historically significant materials. In a period before the advent of air conditioning, operable windows provide fresh air to interior spaces.

A defining characteristic of large libraries in the early 20th century was their controlled access to materials (Kaser, 1997, pp. 107–111). Subsequent

---

[3] While typically traditional in style, there was variation based on location and surroundings, with red brick Georgian and Collegiate Gothic most prevalent.

to their construction, there has been a philosophical shift toward openness and accessibility for collections. It is cumbersome from the patron's viewpoint to fill out call slips and wait for books. It does not suit our contemporary consumer-driven service oriented culture. At the Baker Library, renovations designed by Robert A.M. Stern Architects will make collections more accessible by removing the stack core, while preserving and restoring the original reading room and other major architectural spaces.

While most large academic and public libraries built before the Second World War were designed using the closed stack model, small libraries typically provided direct access to books (Van Slyck, 1995; Breisch, 2001). Carnegie-endowed libraries were conceived as vehicles for popular education and social development; their users were provided with direct access to books as well as to an auditorium for lectures and public meetings. These libraries were more accessible and democratic than earlier libraries, and their architectural design reflected and reinforced this trend.

V. Modular Design in Libraries

After the Second World War, in response to the increased influence of European modernism in architectural design, there was a conscious effort to reconsider the closed-stack model for library buildings. A group of respected librarians, planners, and architects formed the Cooperative Committee on Library Building Plans and met on a regular basis between 1944 and 1948 (Kaser, 1997, pp. 112–113). The Committee published recommendations that had a significant effect on the design community. Principal among these was the replacement of closed stacks with a modular configuration that expanded the universal access model to libraries of all sizes. Book-stacks and patron seating, instead of being segregated, were mixed together. Flexibility was key: the fewer walls and rooms, the better. Modularity allowed for the updating and replacement of building and furniture components over time. Supporting this system would be a regular and often neutral architectural wrapper. This wrapper would provide environmental control: air conditioning systems and a regular grid of lighting fixtures. Air conditioning would allow for large floors without operable windows that lower its efficiency. Flexible space would allow librarians to configure collections in an optimized way. These clear, concise, and logical recommendations, combined with a positive economic climate, led to a new generation of post-war libraries (Kaser, 1997, pp. 113–120).

The advent of the modular library building corresponded with the triumph of the International Style in the architectural community. Leading practitioners and proponents of European modernism, including Walter

Gropius, Marcel Breuer, and Mies Van der Rohe, fled Nazi Germany and came to the United States, where they received high-profile commissions and trained a generation of American architects (Haskell, 1950, pp. 46–47). The technological and socially progressive basis of modernist design, along with its machine-age design aesthetics and its dismissal of historical styles, found resonance among Americans who yearned for a clean start after the horrors of the Second World War. Too often, however, the simplicity and aesthetic elegance of the best modernist design were confounded with cheapness, and the universal nature of the International Style led in many cases to a lack of consideration of local character and conditions.

VI. The Post-modern Reaction

As the memory of the War faded, history reasserted itself. Technology, while still held in high regard, was no longer seen as a panacea. Critiques against the sterility of post-war design began to appear in the 1960s, most notably in the writings of the social critic Jacobs (1992), who wrote *The Death and Life of Great American Cities* in 1961 and architect Venturi (1992), who wrote *Complexity and Contradiction in Architecture* in 1966. Venturi argued that the prevailing modernist approach was too narrow and did not reflect the richness and diversity of society. In his view, architecture is a visual language, and buildings are signs that can be read and deciphered. These critiques ushered in a period of pluralism, in which many approaches to design vied for dominance.

Library design has been influenced by these debates within the architectural community. Among the current approaches is a techno-positivist school that argues that recent developments in computer and information technology are part of a continuum dating back to the industrial revolution. Ubiquitous information technology renders historical forms and architectural traditions irrelevant because they are divorced from contemporary life. The new Seattle Public Library, designed by Dutch architect Rem Koolhaas, is a recent example of this approach.[4] Many architects try to express the dynamism of today's information-saturated society by creating flowing, twisting shapes that seem to dematerialize space.[5] Other designers favor the incorporation of new technology within traditional forms. Many libraries have become hybrids, marrying the flexibility of the modular plan with more traditional reading spaces (Kreyling, 2003, pp. 158–161).

[4]For information about the new library project, see the library's website www.spl.org.
[5]The avant-garde architect Zaha Hadid, one of the leading proponents of this approach, was awarded the Pritzker Prize in 2004 (Muschamp, 2004).

The growing imperative of sustainable design is leading to fresh design solutions that are more resource efficient than those of the recent past.[6]

VII. Today's Library Design Issues

Designing libraries in today's current pluralistic environment is challenging. While history can be a guide, there is no absolute road map to follow. The opportunities presented by new technologies and design approaches should be balanced with the value of institutional tradition and local context. Library design should bridge the "two cultures" of science and literature described by essayist Snow (1961). The following sections discuss some of the significant issues in designing contemporary libraries.

A. Collection Storage

Collection storage makes up the bulk of the space requirement and expense of library buildings. In a traditional library, storing collections is what a library is all about. When architects plan a building, we provide space for current collections and make assumptions about future growth. One of the things that frustrates administrators and those who fund libraries is the inability to accurately and convincingly estimate future collection growth, and hence the optimal size of a library building. Architects usually base their design on growth assumptions prepared by their clients or library planning consultants.

There is a wide range of predictions about the future of collection growth. Some consultants and librarians believe there will be no net growth in space required for collections in the foreseeable future.[7] Their argument is that more and more materials will be available electronically, so libraries can purchase fewer paper copies of materials and also can use inventory management systems to precisely and quickly identify seldom-used materials that should be aggressively weeded, and either deaccessioned or relocated into a remote storage warehouse. By upgrading the stock on the shelves, a library can achieve a more customer-focused orientation and increase usage without adding new building area.

At the other extreme are those who see significant increases in the area required for collections in the future. Their rationale is that while journals may indeed go on-line, there is little indication that monograph publication will decrease in volume and rather, the trend is that monograph publishing

---

[6]For one interesting example, the library at Delft Technical University, see Houber, 2001, pp. 123–134.
[7]Private conversation with library building consultant Jay K. Lucker, September 2003.

has increased year to year. Libraries must also now stock a wide range of non-traditional media to serve their constituents. The net result is that collection space requirements may grow at least as much as in the past. Therefore, significant new areas for collections will be needed over a 30- to 50-year planning cycle.

There are also significant differences in collection growth based on the type of library considered. Research libraries and last copy archives have the mandate to keep increasing collections. These collections will likely increasingly reside in remote storage facilities, despite their inconvenience, as the economics of maintaining large collections and the high cost of prime building space will control the decision-making process. Smaller academic libraries and public libraries, which do not have a research orientation, will likely continue to aggressively weed their collections. Increased automation of circulation records should assist in maintaining appropriate collection size and orientation. The degree of digitization of collections remains a wild card in assessing future collection size.

B. Lighting

Many of the most fervent discussions architects have with clients concern lighting in libraries. At any number of conference programs, librarians decry impractical design solutions: bad lighting is always near the top of the list of problem areas. One particularly interesting program was titled *The Seven Deadly Sins of Public Library Architecture* (Schlipf and Moorman, 1998). To the audience's amusement, the speakers catalogued impractical, stupid, and useless design ideas. These deadly sins included building leak-prone skylights, specifying many different kinds of light bulbs that were hard to re-stock, and placing lights 20 or 30 ft in the air over stairways where they were impossible to reach. Architects, on the other hand, decry clients who ask for beautiful buildings, and then require that the whole building be lit in uniform ranks of 4-ft fluorescent tubes or unwelcoming metal halide lamps.

For reading and writing on paper, natural daylight, properly controlled, provides unparalleled environmental quality. North light has the most uniform level of illumination, and is preferred for reading, as it is for artist's lofts and architect's studios. At night, or when natural daylight is unavailable, it is most pleasant to read with a task light, or lamp, that gives warmth to the page and is localized to shine where it is needed; it is therefore energy efficient. Daylight from the south and west can be too intense for comfortable reading. A variety of shading and screening methods are also available for direct lighting (Olgyay and Olgyay, 1957). Even direct sunlight is often preferable to artificial lighting. Daylight also has the advantage of changing continuously through the day, giving the reader a connection back to the outside world.

Many designers favor indirect lighting, where light reflects off ceiling and wall surfaces rather than directly onto users. An even, ambient light level is created. However, as a result, these spaces can feel cold and static. Indirect lighting is also relatively inefficient in terms of energy use, because the lamps are illuminating surfaces like walls and ceilings, which in turn provide an overall ambient light level within a space.

What truly drives people away from libraries, are the metal halide fixtures and endless runs of fluorescent lights found in many libraries, schools, and offices. These fixtures give off a quality of light that is cold and dehumanizing. If you would not accept this kind of lighting in your home, you should not accept it in places where you and your children are supposed to learn.

Computer screens reflect glare from point sources of light, so the conventional solution has been to place computers in areas with indirect lighting. It is time to reconsider this practice. The computer industry is moving toward laptops and devices with greater mobility, which, coupled with wireless services, will mean that the computer, like the book, will be used in all kinds of light. Laptop screen design has largely overcome the glare problem. Design of lighting for areas with concentrated computer usage should increasingly become similar to other areas within the building. This yields the greatest degree of flexibility as screen technology continues to evolve.

In lighting ranges of stacks, the goal is to create an even intensity of light for each shelf, top to bottom. This is most efficiently accomplished by running lights parallel to rows of stacks, centered on the aisles. Shelving manufacturers have begun to manufacture shelves that project out and tilt at a greater angle as the stacks approach the floor, which will assist in lighting the lower shelves. Architects have been working closely with some manufacturers to promote this ergonomic solution.[8] Some library planners have argued that stack lights should run perpendicular to the shelves, thus providing greater "flexibility" (Thompson, 1977, p. 146). This practice should be curtailed. It is a waste of energy to light the top of the shelves, which cast unwelcoming shadows, and are probably dusty anyway.

---

[8]Ergonomic bookstacks have been created for the Jacksonville Public Library in collaboration between library staff, Robert A.M. Stern Architects, and the manufacturer, Tennsco Incorporated.

C. Furniture

The design of most library furniture is dismal. The library furniture business has historically been a backwater, characterized by ubiquitous boxy wooden chairs with sled legs and boxy wooden carrels with oak finished to resemble plastic laminate. The furniture that pervades many public and academic libraries demonstrates a lack of vision on the part of designers, manufacturers, and specifiers. When looking to the success of bookstores such as Borders and Barnes & Noble in attracting people to stay and read awhile, look first at their furniture.

A few manufacturers dominate the library furniture market. Manufacturers' representatives develop relationships with librarians by providing design and layout services; in return furniture is often specified without consideration of design quality. The most often asked questions are, "How much does it cost?" and "How long is the warranty?" While certainly important, these are by no means the only questions worth asking.

Architects often try to persuade clients to allow them to specify or design the furniture for the library building projects they create. This is because furnishings contribute significantly to the overall look and feel of the library. Librarians undertaking a building project should visit NEOCON,[9] or other major interior design trade fair, rather than restricting themselves to what is presented in catalogs or at the American Library Association convention. They may consider specifying shelving designed for record stores, bookstores, and specialty retailers. Architects and their library clients can work directly with manufacturers with exceptional technical capabilities even if they have not necessarily specialized in the library market. As libraries must increasingly entice users with the quality of their environments, the design quality of library furniture is likely to increase.

D. Flexibility

We cannot predict the future, so we need to hedge our bets. That, in a nutshell, is the argument behind providing flexibility in library plans. Assumptions about collection growth may be incorrect. A new technology may change expectations about what services a library should provide. Service goals are often updated, and new administrators may have new priorities. Construction is expensive, and no one wants to be trapped trying to make do in a building that is outmoded.

---

[9]NEOCON is a large furniture and interior design trade fair held annually at Chicago's Merchandise Mart. It is used by many manufacturers as the venue to introduce new product lines (www.merchandisemart.com/neocon/).

There is often a contradiction between the goal of flexibility, and the desire for architectural quality. Finely crafted spaces are by their nature often static. Yet such so-called inflexible spaces, such as traditional reading rooms, are just the kinds of places that many people love, that draw them back into libraries again and again.

A library should have flexibly designed spaces, but not every space should be flexible. Different areas can be flexible to varying degrees. The nature of collaboration, group study, and the support environment using technology is evolving, and will continue to do so. Libraries should be able to accommodate this evolution. Conversely, the act of sitting and reading a book has not changed in centuries. Libraries should accommodate both of these uses. Other library services reside along the continuum between continuity and change—and the designer's plans should reflect this more nuanced approach to flexibility.

It is also important to remember that some of the best renovations are unexpected. For example, when I was working on a project in Paris several years ago, our French associates had their office in a restored 18th century townhouse built around a courtyard that had at one time been a blacksmith's shop. At the time when it was built, no one imagined that it would become home to a company of over 50 architects, let alone with electric lights and computers. Yet the space functioned admirably, and offered a warmth and charm that a purpose-built building would have lacked. Successful libraries have been designed into remodeled houses, shopping centers, and in the case of the recently completed Yonkers Public Library, a former Otis Elevator factory, a building that had most recently been used to warehouse locomotive parts.[10] Architects and their clients should take a broad view of what is considered flexible space.

E. Sustainable Design

Buildings use tremendous levels of natural resources, both in their construction and in their operation. Architects, contractors, and clients have undertaken the massive task of increasing the efficiency and sustainability of their designs, but progress has been agonizingly slow. Unlike other industries that have a few dominant leaders, the building industry is extremely diffuse, with a complex network of designers, specifiers, producers, distributors, and contractors, as well as a seemingly limitless range of choices. Many building firms and product manufacturers are small and undercapitalized, so research and development is nearly non-existent.

---

[10]Robert A.M. Stern Architects provided the design concept for the Yonkers Public Library and Board of Education; see "Governor Joins Celebration at Opening of Larkin Center," press release (www.state.ny.us/governor/press/year02/sept18_02.htm).

Designers often rely on unconfirmed and sometimes self-serving product information provided by manufacturers.[11]

Nonetheless, progress is being made. There are two fundamental ways to lessen the environmental burden of library buildings: fewer resources can be used in their construction, and fewer resources can be used in operations and maintenance (US Green Building Council, 2003). Both of these are affected by the building design. In the early planning stages of a project, it is worth reviewing the minimum required construction needed to accomplish institutional objectives. Is a new building necessary, or will a renovation of existing space accomplish nearly the same result? Is there other underutilized space available that can be upgraded? Both library programmers and architects tend to have an expansive view of the library's institutional mission, but from an environmental perspective, building less is often the preferred solution.

A related goal is to build for the long term. Nothing is more destructive than tearing down a building after 20 years because it was cheaply made, was not adaptable to change, or was merely ugly and unloved. An institutional project like a library should be conceived to last 100 years with periodic renovation to mechanical systems and interior furnishings. Many buildings of quality have lasted much longer.

In order to build more environmentally sustainable libraries, it is important to understand the sequence of manufacture that precedes the arrival of a building material onto a construction site (McDonough and Braungart, 2002). For example, most carpets are oil based; wallboard relies on the mining of gypsum; many beautiful wood veneers are associated with clear-cutting of primary growth forests; the production of steel creates air pollution; and aluminum manufacture is a significant user of energy. Many materials are trucked and shipped across the country and even around the world. Most construction decisions are cost based, which often obscures their environmental impact. As a general rule, locally produced materials, simple manufacturing methods, and renewable materials produce the least environmentally damaging buildings (US Green Building Council, 2003, pp. 221–227).

Ongoing maintenance also is a significant resource burden. Mechanical and lighting systems use large amounts of energy. Designers can lower the resulting burden by using the building itself as a collector and distributor of light and energy (Lamis, 2003, pp. 36–40). As one example, walls and ceilings can store up solar energy during the day and re-radiate at night, lowering the need for auxiliary heat. Technically unsophisticated approaches to the movement of water and air through buildings, along with proper orientation to take best advantage of the position of the sun and prevailing winds, can

---

[11]For one of the most objective sources for sustainable building materials, see GreenSpec Directory.

lower energy use significantly. American buildings are particularly wasteful in their use of energy; buildings in Europe often use a fraction of the energy to perform similar functional purposes. We can learn a lot from the work of traditional builders who met their needs with much fewer resources than we use today (Rudofsky, 1964). More contemporary approaches can also help us, including our ability to pinpoint control over our environment. For example, photo-sensors can tell us how much natural light is coming into a building and adjust the levels of the artificial lighting accordingly.

Libraries have a high social purpose, and should be inspirational leaders in the drive for energy and resource efficient building designs.

VIII. Conclusion

To paraphrase Dickens, it is the best of times and the worst of times to be a library designer. Library design is in a state of flux; and many old truisms may no longer apply. Libraries face significant challenges to defining their mission. Are libraries as institutions tied to library buildings? Increasingly the answer will be no. Libraries are no longer the storehouses of knowledge that they were through most of their history. Knowledge started leaking out of library buildings around 1980 and now is pouring onto the Web. Information is everywhere. People no longer *need* to go to library buildings; they must *want* to go. This puts those of us whose interest is bound up with the success of library buildings in the position of trying to figure out what people want, rather than telling them what is good for them.

So what do people want? They certainly want convenience. They want access to information and they want it now. They want comfort. The sterile unadorned libraries that were built on college campuses and in cities around the country from the 1950s through the 1980s cannot compete with coffee shops like Starbucks, bookstores like Barnes & Noble, or your home.

People who go to an academic or public library want to feel that they are a part of something important, a tradition. This association is an essential part of the library's hold on the public imagination, and while it may have little do with functional layout or efficiency, it has everything to do with enticing people to come in the door and stay.

The Internet and various electronic media have had a strong centrifugal effect on contemporary culture and increasingly, the built environment. Information can come to you wherever you are. However, human nature is still social—people want to be with others. There is also a great store of goodwill and passion for the "library as place".[12] In a knowledge-driven

---

[12]For a recent analysis of library building projects, see Shill and Tonner, 2003, 2004.

society, libraries can be our knowledge-based community and campus centers. But there is no standing still: libraries must move forward to align themselves with current conditions, or they will retreat into a marginal role as gatekeepers to an increasingly musty past.

References

Albanese, A. R. (2001). Libraries as equity building blocks. *Library Journal*, 40–42. May 15.
Baur-Heinhold, M. (1972). *Schöne alte Bibliotheken: Ein Buch vom Zauber ihrer Räume.* Verlag Georg D.W. Callwey, Munich.
Breisch, K. A. (1997). *Henry Hobson Richardson and the Small Public Library in America.* MIT Press, Cambridge, MA.
Cruikshank, J. L. (1987). *A Delicate Experiment: The Harvard Business School 1908–1945.* Harvard Business School Press, Boston, MA.
de Ménil, D. (ed.) (2002). *Visionary Architects: Boullée, Ledoux, Lequeu; Exhibition Catalog.* Hennessey + Ingalls, Santa Monica.
Dubbin, R. (ed.) (2002). *Henri Labrouste.* Elemond Spa, Milan.
Garreau, J. (1991). *Edge City: Life on the New Frontier.* Doubleday, New York, NY.
GreenSpec Directory, 4th ed. Building Green Inc. See www.buildinggreen.com.
Haskell, D. (1950). Gropius' influence in America. *L'architecture d'aujord'hui*, 46–47. February.
Houber, F. (2001). *Mecanoo Architects: Composition, Contrast, Complexity.* Birkhäuser, Basel.
Jacobs, J. (1992). *The Death and Life of Great American Cities.* Vintage Books, New York, NY.
Kaser, D. (1997). *The Evolution of the American Academic Library Building.* Scarecrow Press, Lanham, MD.
Kreyling, K. (2003). Nashville public library. *Architectural Record*, 158–161. February.
Kunstler, J. H. (1993). *The Geography of Nowhere.* Simon & Schuster, New York, NY.
Lamis, A. P. (2003). In *Greening the library: an overview of sustainable design* (G. B., McCabe and J. R., Kennedy eds.), pp. 36–40, Planning the Modern Public Library Building, Libraries Unlimited, Westport, CT.
Le Comte de Laborde (1845). *De L'Organisation des Bibliothèques Dans Paris.* A. Franck, Paris.
Leopoldo della Santa (1816). *Della Costruzione e del regolamento di una publica universale biblioteca con la pianta dimostrativa.* Firenze. Referenced in: Le Comte de Laborde, *De L'Organisation des Bibliothèques Dans Paris* (1845). A. Franck, Paris.
Lerner, F. A. (1998). *The Story of Libraries: From the Invention of Writing to the Computer Age.* Continuum, New York, NY.
McDonough, W., and Braungart, M. (2002). *Cradle to Cradle: Remaking the Way we Make Things.* North Point Press, New York, NY.
Mitchell, W. (1995). *City of Bits.* MIT Press, Cambridge, MA.
Moore, C. (1970). *The Life and Times of Charles Follen McKim.* DeCapo Press, New York, NY, reprint of 1929 original published by Houghton Mifflin Company, New York.
Muschamp, H. (2004). An Iraqi-born woman wins Pritzker architecture award. *New York Times* March 22.

Negroponte, N. (1995). *Being Digital*. Vintage Books, New York, NY.
Olgyay, V., and Olgyay, A. (1957). *Solar Control and Shading Devices*. Princeton University Press, Princeton, NJ.
Richards, J. M. (ed.) (1977). *Who's Who in Architecture from 1400 to the Present*. Holt, Rinehart and Winston, New York, NY.
Ripley, C. (2003). The power of libraries. *Urban Land*, 96–98. October.
Rudofsky, B. (1964). *Architecture Without Architects: A Short Introduction to Non-pedigreed Architecture*. Museum of Modern Art, New York, NY, Distributed by Doubleday, Garden City, NY.
Schlipf, F. A., and Moorman, J. A. (1998). *The Seven Deadly Sins of Public Library Architecture*, program given on March 12, 1998 at Public Library Association Convention, Kansas City. Information available at www.urbanfreelibrary.org/fredarc.htm.
Shill, H. B., and Tonner, S. (2003). Creating a better place: physical improvements in academic libraries, 1995–2002. *College and Research Libraries* **64**(6), 431–466.
Shill, H. B., and Tonner, S. (2004). Does the building still matter? Usage patterns in new, expanded, and renovated libraries, 1995–2002. *College and Research Libraries* **65**(2), 123–150.
Snow, C. P. (1961). *The Two Cultures and the Scientific Revolution, The Rede Lecture, 1959*. Cambridge University Press, New York, NY.
The Better Jacksonville Plan Website. www.betterjax.com.
Thompson, A. (1963). *Library Buildings of Britain and Europe*. Butterworths, London.
Thompson, G. (1977). *Planning and Design of Library Buildings*. The Architectural Press Ltd, London.
US Green Building Council (2003). *LEED Reference Guide for New Construction and Major Renovations*. US Green Building Council, Washington, DC, see www.usgbc.com.
Van Slyck, A. A. (1995). *Free to All*. The University of Chicago Press, Chicago.
Venturi, R. (1992). *Complexity and Contradiction in Architecture*. Museum of Modern Art, New York, NY.

Excellent Libraries: A Quality Assurance Perspective

Felicity McGregor
University of Wollongong Library, Wollongong, NSW, Australia

I. Introduction

A. Background

The proliferation of inspirational leadership and management publications available in libraries and bookshops suggests that there are many paths to excellence. Much of the literature is written with a business or corporate audience in mind; however, it is a source of ideas, theories and models that, potentially, can be applied in public or not-for-profit organisations. One theory which has enjoyed a long history of debate and discussion in management studies is quality management, variously referred to as TQM, quality assurance, total quality control or one of the many other alternatives. In this chapter the applicability and potential benefits, as well as the challenges and obstacles, of adopting one version of total quality management in a library setting are examined.

This discussion of the application of quality management in libraries is based on the experience of the University of Wollongong Library (UWL) in selecting and adopting the Australian Business Excellence Framework (ABEF), administered by Standards Australia International.[1] In adopting a quality framework, hereafter referred to as the ABEF, UWL intended to evaluate its progress towards its stated vision, mission and goals by applying for the associated Australian Business Excellence Award (ABEA). The latter includes a major submission and rigorous on-site audit by qualified evaluators. Organisations can choose to enter the awards at different levels.

---

[1]The Australian Business Excellence Framework was previously known as the Australian Quality Framework and was administered by the Australian Quality Council. Similarly, the Australian Business Excellence Award was previously known as the Australian Quality Award.

In 1996, less than 2 years after adopting the framework, UWL was evaluated and received recognition at "achievement" level. Two years later, evaluation at Award level resulted in reaching Finalist status and in 2000, UWL became the first library to compete with a range of profit and not-for-profit organisations to receive an Australian Business Excellence Award.

The "quality journey," as it is commonly known, provided the opportunity to examine all elements of the Library, its structures, systems, services, processes and people. Through a lengthy process of planning, implementation, review and improvement, the goals of the quality program were achieved. Of greater significance, the organisational learning and development which was integral to the journey was more far-reaching and transformational than UWL leaders could have envisaged at the outset.

Reflection on both progress and process has been a feature of the journey. The adoption of what was widely perceived as a business-oriented management system was new in the library world and attracted interest from both within and outside the profession. The process of internal and external reflection produced insights which may be of interest to others and are recorded below, chiefly within the context of each section heading, as well as in Section VI, "Challenges and Insights".

Although the role of leadership is not discussed in detail, without the vision, commitment and perseverance of leadership, transformational changes of the kind described in this chapter are unlikely to be wholly successful. Intrinsic motivation is briefly discussed in the conclusion; it is a critical driver for leaders and change agents. Possessing the relevant competencies for change management, discussed in Section VI is vital; *wanting* to transform an organisation, whatever it takes, is an irreplaceable leadership attribute.

B. Context

The UWL is the primary information and resource service for the University of Wollongong, a medium-sized university, located in the Illawarra region 80 km south of Sydney, NSW. The Library provides services and resources to the central campus, a South Coast campus and access centres, the Sydney Business School, a campus in Dubai, the Wollongong University College (entrance level students) and a network of geographically remote students. Services are also provided to the University's business arm, UniAdvice, to strategic partners, such as local industry and to alumni. Although our primary clientele are those connected with the University, we provide services to the Illawarra community wherever possible.

In 1975, the University was established as an independent entity. The Library at that time was a small traditional, inwardly focused operation comprising 99,415 volumes, 1100 subscriptions and a clientele of approximately 2000 students, mainly local high school graduates, qualified for university entrance. Today, the Library has expanded its range of services to match the growth of the University and is considered a leader in the higher education sector in continuing to deliver quality services during unprecedented changes in Australian universities and during a similarly transformational period for library and information services.

The demand for a traditional book-based service, centred on a single location, is still a significant component of our business and, while there is a slight downward trend in clients entering the building, 11% over the last 3 years, loans have increased in the same period by 13%. It is too soon to consider that the print-and-study-space concept is obsolete. The physical space provided in 1975 approximated 4600 m^2. Today, available space is over 10,000 m^2, which includes the Shoalhaven Campus and the Curriculum Resources Centre. The collection includes 660,000 volumes, and access to 21,000 serials, mostly in electronic format. These resources, as well as electronic books and readings reflect the needs and expectations of clients for resources and services which can be delivered independently of the physical location of either the service provider or the consumer of the service.

Undergraduates now number approximately 18,000 including those in remote locations. Over 50% of students, including the 25% who are international students, are drawn from outside the region. Other major client segments include 969 academic staff and 4390 postgraduate students. The Library receives 82% of its budget from the University's operating grant, 11.3% from international fee-paying students and 6.7% is earned as income.

II. The Search for Excellence

In its earliest days, the Library was characteristic of its time in being conservative, hierarchically structured and risk-averse. The appointment of a new University Librarian in 1986 signalled an agenda for change. Strategic planning was instigated in that year, accompanied by management training for senior staff. Organisation-wide staff development was embraced as a driver of change and a staff development committee was formed in 1988. With the establishment of the new position of Deputy University Librarian in 1989, it was possible to accelerate the pace of change. From 1989 to 1994, experimentation with various management models resulted in a team-based structure, increased involvement of staff in planning, progress in the use of technology and a performance management system. These developments formed a useful foundation on which to build further improvement.

As the pace of technological change accelerated and expectations of services increased, it became apparent that change would be a constant, but often unpredictable, factor in the library and information environment. The Library executive recognised that an appropriate framework in which to manage constant change in all aspects of the business would be critical to future success and sustainability. Following consideration of a number of factors, the ABEF, see note 1, was selected in 1994 as the Library's management and change-guidance framework.

The Framework provided a structured and integrated management system. It linked a focus on people and clients, leadership and planning, areas which had been progressively improved, with data and information systems, process management and improvement, and an emphasis on business results, areas which had been less rigorously addressed. It appeared possible to commit fully to the principles underpinning the framework as they accorded with existing management values and with the business philosophy the Library executive hoped to adopt in the future.

The Library commenced what is commonly known as a "quality journey" in 1994. There were many reasons for the choice of the ABEF to guide this "journey." Questions uppermost at the time, and perhaps worthy of contemplation today were: How does a rather small, sparsely funded, regional university library aspire to excellence? How does a library established in 1975 compete, in an increasingly competitive environment, with wealthy, centuries old, metropolitan libraries?

In the library world, excellence has traditionally denoted extensive collections, capacious facilities, sufficient staff and, yes, a service orientation. The first three attributes are largely resource dependent. In the provision of service, however, there appeared to be a clear opportunity to excel. Excellent service, it was believed, required excellent people and high-quality, cost-effective supporting processes.

> While acknowledging that, in many instances, there is no substitute for significant, comprehensive on-site collections, technologically driven improvements in the distribution of, and access to, resources has seriously undermined the *bigger is better* value proposition. The perception that wanted information is ubiquitously and freely available and that libraries no longer have a vital role in universities has provided further impetus, in terms of future viability, for libraries to demonstrate that they are not only essential to the success of the university's researchers and students, but are of strategic importance in achieving the university's mission and goals. (McGregor, 2000)

In adopting the ABEF, discussed in more detail below, the Library not only sought to improve its performance in all areas but to measure and compare performance and outcomes with other organisations. The result was to demonstrate competitiveness with others and the primacy of the Library's role in assisting the University to achieve its goals, especially those aimed at the attraction and retention of students.

The banner chosen for the introduction of the program was *Quality Service Excellence*. These three elements underpin most quality philosophies and signalled succinctly the key aims of the UWL quality program.

III. Defining Quality

There is no shortage of literature on the subject of quality management. The terminology includes quality assurance, continuous improvement, total quality service and total quality management. Groenewegen and Lim (1995) discuss some of the definitions of TQM, its use in libraries and universities and the interpretation of quality in these contexts.

"Quality assurance" tends to be associated with industry and implies an emphasis on procedures and documentation. As Dawson and Palmer (1995, pp. 14–15) explain, "…QA operates by the use of documented formalised procedures which can be monitored and evaluated by internal QA inspectors and assessed by external quality agents for local, national and international accreditation."

"Quality" is a prevailing, if poorly defined, concept in universities. "…the literature on quality in higher education is scattered with assumptions that a university is about quality." (Groenewegen and Lim, 1995, p. 6). With the establishment of the Australian Universities Quality Agency (AUQA) by the Federal Government in 2000, "quality assurance" and "quality audit" have greater currency in Australian universities. However, the understanding of these terms by AUQA appears to be closer to the meaning ascribed to TQM than to QA.

Quality audit is defined by AUQA as "a systematic and independent examination to determine whether activities and related results comply with planned arrangements and whether these arrangements are implemented effectively and are suitable to achieve objectives" (Australian/New Zealand Standard, 1994 quoted in the AUQA Audit Manual Version 1 May 2002).

The manual goes on to explain that the purpose of audits is "to investigate the rigour and effectiveness of the organisation's performance monitoring against its plans" and that relevant processes and mechanisms are "effective in achieving the stated goals" (AUQA Manual, 2002, p. 17).

Without re-examining these definitions or listing the many concepts attributed to quality, the definition which best describes the understanding and the aims developed for the UWL quality program is: "TQM is defined as a structural system for creating organization-wide participation in the planning and implementation of a continuous improvement process that meets or exceeds the expectations of the organization's customers or clients. As many organizational development experts have noted, 'TQM is a journey, not a destination'" (Shaughnessy, 1995, p. 1).

Shaughnessy goes on to discuss the problems which the TQM terminology may cause with its focus on management, frequently an unpopular term in universities. Because of the emphasis of TQM on process improvement, it is sometimes argued that the system pays insufficient attention to outcomes. However, "an overriding objective of TQM is the improvement of the quality of customer outcomes…" (Shaughnessy, 1995, p. 2).

In 1994 when the Library embarked on its "quality journey," AUQA had not been established. The desire to be audited or evaluated against recognised standards, however, was important to the Library then and remains important today. Although library services are included in an AUQA audit, given the scope and complexity of universities' core purposes, teaching, research and learning, audit schedules generally do not permit a detailed investigation of the library and other supporting elements of the university. Moreover, AUQA's 5-year audit cycle is an unacceptably long gap when assessing progress in a rapidly changing environment.

The framework adopted by the Library is now known as a business excellence rather than a quality framework.[1] Interestingly, the reason for the change was the unpopularity of the "quality" terminology. For the Library of 1994, the "business excellence" label would have presented a barrier in terms of acceptability. By 2000, when the Library applied for a business excellence award, the terminology was no longer an issue.

IV. Adopting a Business Excellence Framework

Prominent in the 1994 decision-making process was the desire for a total management framework to guide the implementation of improvements to internal structures, systems and processes. Investigation suggested that such a framework would also assist in identifying further improvements as well as enabling effective management of the inevitable changes and developments mandated by external forces, such as the revolution in information and communication technologies.

A quality or business excellence framework was seen as a model for organising and integrating initiatives and building on previous change interventions. McGregor (1991) describes an early change intervention at UWL. At this time, the obstacles to change were considerable and included staff resistance, low morale, low performance, limited distribution of managerial skills and limited commitment to organisational growth and improvement.

A key learning from this earlier change effort was that the involvement of staff in planning and implementing change is critical. Although accepted as a truism in change management theory, implementation of the theory through genuine involvement in planning, as opposed to the mere communication of information about plans, is probably less well accepted and practised. Another insight was the need for extensive staff preparation. Skill development and educational opportunities for all staff are vital facilitating factors in any change effort. Often the training and development associated with change implementation is directed to supervisors or group leaders alone.

One of the "Principles of Business Excellence" on which the ABEF is premised is: "the potential of an organisation is realised through its people's enthusiasm, resourcefulness and participation." (ABEF, 2003). This principle was already internalised by the library executive as evidenced by the commitment to staff development and performance management established prior to 1994.

All of the 12 principles (see Table I) were philosophically acceptable to the Library executive and, in essence, they comprise statements of good management practice. As stated in the Framework: "These foundational Principles, which have evolved over the past 50 years, are supported by a body of published research that underpins all similar frameworks throughout the world" (ABEF, 2003).

Of considerable appeal, therefore, was the ABEF's utility as a holistic management framework. The disparate elements of effective management practice: human resources, industrial relations, customer relationship management, leadership strategies and planning processes are all integrated in a model underpinned by a systems approach and informed by systematic data collection, information and knowledge management (see Fig. 1).

As reported in McGregor (2003), the seven categories create a specific structure or context in which organisations can review, question and analyse their leadership and management system.

The *Leadership & Innovation* and *Customer & Market Focus* categories are seen as *drivers* of all other components. The *Strategy & Planning Processes* and *People* categories are shaped by the drivers and can be seen as supporting processes that *enable* or facilitate achievement in all other areas. The *Data, Information & Knowledge* category is shown as weaving throughout the model to illustrate its integration across all aspects of the organisation.

The *Processes, Products & Services* category is shaped by the drivers, supported by the enablers and fundamentally focused on how work is done to achieve the required results of the organisations. The *Business Results* category is about organisational outcomes or overall performance and depends on the design of, and interrelationship between, the other six categories. If organisations want to change their *Business Results*, then they must improve

Table I
The 12 Principles of Business Excellence

(1) *Direction.* Clear direction allows organisational alignment and a focus on the achievement of goals
(2) *Planning.* Mutually agreed plans translate organisational direction into actions
(3) *Customers.* Understanding what clients value, now and in the future, influences organisational direction, strategy and action
(4) *Processes.* To improve the outcome, improve the system and its associated processes
(5) *People.* The potential of an organisation is realised through its people's enthusiasm, resourcefulness and participation
(6) *Learning.* Continual improvement and innovation depend on continual learning
(7) *Systems.* All people work in a system; outcomes are improved when people work on the system
(8) *Data.* Effective use of facts, data and knowledge leads to improved decisions
(9) *Variation.* All systems and processes exhibit variability, which impacts on predictability and performance
(10) *Community.* Organisations provide value to the community through their actions to ensure a clean, safe, fair and prosperous society
(11) *Stakeholders.* Sustainability is determined by an organisation's ability to create and deliver value for all stakeholders
(12) *Leadership.* Senior leadership's constant role-modelling of these principles and their creation of a supportive environment to live these principles are necessary for the organisation to reach its true potential

in all six categories (summarised from Standards Australia International, Australian Business Excellence Framework, 2003). For a more detailed description of the framework, its categories and items, see the Framework document and papers by McGregor (1997, 2000, 2003).

As well as achievements in the "People" category, the Library had a history of strategic planning, a track record as early adopters of technology and a focus on service and user assistance commonplace in so many libraries. In the categories of "Strategy & Planning Processes" and "Customer & Marketing Focus," there were improvements to make but the foundations were present. Other categories, however, presented challenges of greater magnitude.

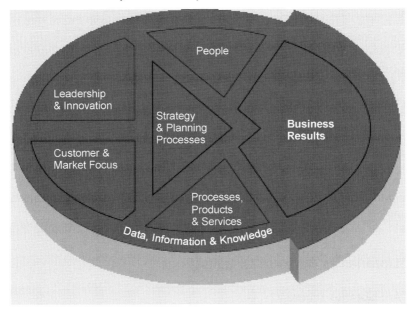

Fig. 1 Australian Business Excellence Framework (ABEF, 2003, p. 13).

The "Data, Information & Knowledge" category involves collecting, analysing and presenting data to use in prediction, performance measurement and decision making. In other words, input and output measures such as budget quantum, collection growth, circulation, reference enquiries and so on would be insufficient to meet the requirements of this category.

Similarly, the "Business Results" category with its emphasis on indicators of success and sustainability presented a daunting hurdle. In spite of the immensity of the challenges inherent in these and other categories, it was concluded that they were not insurmountable. It was recognised that an increasingly complicated, constantly changing library and higher education environment required a management structure to assist in clarifying and managing the complexity. On reflection, the framework has been invaluable in maintaining a holistic perspective and in reducing preoccupation with technical and technological problems. The emphasis on establishing direction and long-term goals is an antidote to the distraction of immediate problems.

As mentioned above, for reasons of competitiveness with other information providers and for positioning within the university, the Library sought benchmarks or performance indicators that would measure outcomes and overall organisational performance. These indicators would, desirably, be sufficiently robust to withstand close scrutiny by stakeholders and would enable the Library to compare itself with recognised "best practice" organisations. Within the Library and information sector at the time, benchmarks and indicators to measure effectiveness as well as outcomes were difficult to find.

In adopting the ABEF, there was a clear commitment to evaluate overall performance through application for a quality or business excellence award. UWL was one of the first to enter this arena and the first to be successful in achieving an Australian Business Excellence Award. Other award-winning organisations include a wide range of corporate organisations, legal firms, public service organisations and utilities and local government. Size varies from small, local firms to multinational corporations. Inevitably, in the decision-making process, the question of the relevance of a business framework to libraries was canvassed at length.

V. Libraries as Business Organisations

A. Are Libraries Different?

Business organisations primarily measure the return on their investment, their profit margins and, for some, the return to shareholders. Businesses produce goods or services that are sold for profit. Libraries are different in that their main product, information, is not "used up" when "consumed" and does not usually generate a cash flow. They are "public good" organisations.

The value or return on investment delivered by libraries is of a social, educational, or cultural value and this is difficult to measure. The difficulty of applying accounting standards designed for commercial enterprises is discussed in Carnegie and West (2003). Difficult as it is to measure the value of library resources and services, since they are funded mainly by the taxpayer's dollar, it is reasonable to expect to demonstrate some accountability or return on the funding body's investment.

If, however, libraries are compared with other organisations in terms of functions and structures, then the differences between profit and not-for-profit organisations appear, at least on the surface, to be minimal. Like corporations, libraries are required to manage budgets, and may generate a surplus, if not a profit.

Leadership, strategic planning and human resource management are as essential to effective and efficient libraries as they are to good business organisations. Their importance is reflected in the growth of management programs for librarians, emphasis on management education in library schools and the growth of journal literature devoted to these topics. Like other organisations, the impact of technological change has been pre-eminent in the last 20 years and management of information and communication technology consumes a large proportion of library leaders' portfolios. In recent years, concepts such as client relationship management, partnership management and promotion and marketing have assumed greater importance in libraries, as they have in the commercial world.

Perhaps one of the most controversial issues, especially in higher education, is whether the adoption of business principles and practices should extend to thinking of, and referring to, students as customers.

B. Are Students Customers?

A number of articles have addressed the concerns some academic staff and university administrators have with using this terminology in the education environment. Schwartzman (1995), in considering the application of TQM to education, discusses the advantages of portraying students as customers such as recognising them "as participants in the educational process instead of passive recipients of whatever the institution decides to dish out." He concludes, however, that the advantages are "outweighed by the dissimilarities between commercial transactions and education" (Schwartzman, 1995, p. 1).

Quinn (1997), in discussing the application of quality concepts in the non-commercial setting of academic libraries, sees difficulty in defining the customer in an environment of many different potential customers such as students, faculty, administrators and parents, all of whom may have different expectations of the Library.

This is a dilemma posed to many organisations serving a diverse customer base and is not confined to academic libraries. It can be overcome, however, through the process of developing performance indicators. The articulation of all customer and stakeholder groups and their unique needs and expectations helps to firstly identify the various customer segments the library must serve, and secondly to signal the performance areas that are of vested interest to the various customer groups, for example, see Table II. The UWL uses the terminology "client" rather than "customer." The Performance Indicator Framework (PIF) lists all of the performance indicators which have been identified as relevant guides or gauges of performance against goals and critical success factors. The PIF also lists some of the measures against each indicator. These are the actual data-collecting methods. A key instrument used for measuring client satisfaction is the Rodski Customer Survey which has been adopted by all Australian university libraries. It has similarities with the North American instrument, LibQual + .

Table II
Extract from UWL Performance Indicator Framework

Client group	Expectations	Performance indicators	Measures
Students	Service excellence; knowledge and understanding of needs; skills to identify, locate and evaluate information; access to resources and facilities; collections relevant to their needs	Access to resources	% Materials immediately available; shelving accuracy; database usage
		Client satisfaction	% Clients satisfied (Rodski Customer Survey); number and type of client feedback incidents
		Access to information literacy	Number of clients participating in information literacy tuition; workshop evaluations
		Facilities use rate	Facilities use; entry gate counts
University executive	Leadership in the library and information community; satisfaction of the scholarly information needs of the University; expertise in the navigation of complex and diverse scholarly information environments; cost-efficient operation	Leadership effectiveness	Number of staff involved in professional committees; % strategic plans achieved; benchmarked leadership results
		Effective budget utilisation	Expenditure against targets; processing costs
		Client/stakeholder satisfaction	% Clients satisfied; number and type of client feedback incidents; number of clients using services
		Information Resources Fund usage	Expenditure against targets; cost of supply; speed of supply; collection relevance

The arguments against the "customer" terminology advanced by Quinn and others are more than adequately dealt with in the work of Hernon et al. (1999). Sirkin puts the case simply: "A library patron or user is a customer. He or she is demanding a service and expects that service." (Sirkin, 1993, p. 72).

The views of many of the Wollongong faculty would echo those of Schwartzman and Quinn. Although well aware of these views and addressing them with diplomacy, UWL staff and their student customers have, nevertheless, found the customer service concept empowering and fulfilling. As one UWL client said via the feedback system: "I don't know what you guys have done in this quality management stuff, but it shows!"

In Australia, with an increasing number of students paying either a percentage or the whole of their tuition fees, the demand for excellent customer service seems likely to rise to a crescendo. A recent example of the changing perspectives of students is found in a newspaper article describing a strike by university staff in Sydney: "When I'm paying that amount of money, I'm a customer and I want to be treated like a customer" (Australian Higher Education Supplement, 2003). The student had paid $20,000 up front for his degree and said the strike had cost him a couple of hundred dollars.

In concentrating intensively on client service satisfaction, UWL recognised that excellence would not be achieved by considering service satisfaction alone or in isolation from other factors directly contributing to both satisfaction and quality. Hernon and Whitman (2001) and Hernon et al. (1999) have extensively explored the relationships between customer satisfaction and service quality. In applying the ABEF, the Library was able to embrace and to measure both service quality and client satisfaction. The salient feature of the Framework is that all aspects of organisational management and development are interrelated. It is "an integrated leadership and management system that describes elements essential to organizational excellence" (ABEF, 2003, p. 5). Further, there was a strong belief that staff skills, knowledge and attitudes, as well as their satisfaction levels, had a direct impact on the quality of service.

Heskett et al. (1997) make the case for a strong correlation between customer satisfaction and employee satisfaction. Hernon and Whitman (2001) quote from Heskett and go on to say: "The attitude and role of staff members are key to any service organization that values its customers. While the library has no choice over whom its customers are, the library does control the selection of employees. For this reason, it makes sense to hire staff who have a customer service interest, indeed fervor; to train them accordingly; and to equip them with the authority to satisfy the customer within the context of the vision and mission of the library." (Hernon and Whitman, 2001, p. 39).

Client service attributes became key criteria in position descriptions at UWL, were included in advertisements and interviews and were reiterated during the induction process. From the outset, potential employees were made aware of the importance of the Library's vision for quality, service and excellence. Equally important was the development of the desired attributes in all existing staff, not just new staff or those involved in frontline services. To quote Heskett "In many services, satisfaction is mirrored in the faces of customers and the people who serve them...but it's clear that this magical interaction doesn't occur without a great deal of preparation and thought." (Heskett *et al.*, 1997, p. 111).

In hiring and training all staff, regardless of whether their primary location would be technical services or frontline services, customer service skills were deemed to be essential criteria for selection. All Library staff participated, and continue to participate in staffing service desks. In this way, focus on clients and their needs are maintained, as is the awareness that excellent service is central to the mission, values and performance of the Library.

It is sometimes asserted that library staff have an inbuilt service ethic and that no further attention to this attribute is needed. Regardless of predisposition or personality, consistently excellent service means organisation-wide commitment on the part of all staff. Client focus, therefore, is enshrined at UWL in values, client charters, service standards, policies and position profiles. Training and development opportunities are provided for all staff, including casuals, and programs are updated regularly. Feedback from clients is solicited, welcomed, responded to within a standard time period and acted upon, wherever possible.

Excellence in service requires constant attention and reinforcement. In 1995, the second year of the "quality journey," a slogan *Year of the Client* was adopted to reaffirm the Library's commitment to excellence in this critical area. Actions that year included formation of a Client Service Committee, a major Client Survey, development of training programs in client service and formation of a client service quality improvement team. All staff signed off on a commitment to client service incorporated in a booklet of service guidelines. The guidelines were developed in a series of workshops in which all staff participated and are now included in induction kits for new staff.

The Client Service Committee led the achievement of a number of objectives including the development of service standards and the introduction of a feedback form to capture compliments, comments and complaints (CCCs). This formed the foundation for a systematic approach to client feedback. Now available online and supported by a database, the "CCCs" have provided a wealth of useful data over time and generated many improvements to all aspects of services.

Recognition and reward strategies were also addressed early in the journey and the Client Service Committee drafted the criteria for a client service award that has been awarded bi-annually since 1995. A client service policy, developed in the same year, outlined the elements of service guaranteed to clients. These include reliability, consistency, quality, courteous staff and a safe and clean environment suitable for study.

Enhancing client focus meant paying attention to all of the multifarious components outlined above. This, however, did not constitute a major obstacle to the introduction of a change strategy such as that mandated by the ABEF. In introducing other components of the Framework, the Library found that while some staff were eager to embrace and lead the changes, others were reluctant and slow to participate. An explanation may be found in the perceived characteristics of library staff. Alternatively, as change theories suggest, any organisational grouping will include a percentage of people who embrace change, another group which accepts it and usually, a smaller percentage of people who actively resist change. A brief consideration of how library staff may differ from other employee groups follows.

C. Are Library Staff Different?

Perceptions of professions and occupations tend to be based on stereotypes. It is not the purpose of this chapter to investigate this phenomenon. In an early piece of research in library school, this author found that the stereotype of the mousy, dowdy, almost invariably female librarian was the norm in a range of popular and more serious literature. An Australian newspaper article of the 1980s described librarianship as the "grey blur" profession. Perhaps perceptions have changed since then, as librarians are increasingly recognised for their early adoption and expertise in technology-based innovation, for their involvement in teaching and for possession of a range of skills vital to the information society.

Perceptions seem to persist, however, and personality inventories, such as the Myers–Briggs Type Indicator® (MBTI), to some extent, give credibility to aspects of the stereotype. Research conducted by Myers *et al.* (1998) in their investigation of type theory supports the claim that certain MBTI types appear to be more attracted to some occupations than others. Myers' research finds that "…the modal type of librarians is ISTJ" (Myers *et al.*, 1998, p. 301).

ISTJ is the descriptor for personality preferences for "introversion," "sensing," "thinking" and "judging." "ISTJs appear to be attracted to, and are probably comfortable in, work environments that are efficient, secure, predictable, and conservative, and that permit and promote personal responsibility in their work lives" (Myers *et al.*, 1998, p. 66). The authors go on to say that research suggests "stability and personal control in the workplace are increasingly rare. The qualities that are valued, such as teamwork, rapid adaptability to change, flexibility and the like, are not typically comfortable and natural parts of the ISTJ personality" (Myers *et al.*, 1998, p. 67).

The following is a very brief description of the ISTJ preferences and their associated characteristics: "Introverted: Concentrate quietly on ideas and information; Sensing: Look at facts; Thinking: Analyse information objectively and Judging: Follow an organized system to find materials" (Myers *et al.*, 1998, p. 301). These characteristics fit well with the traditional cataloguing role, for example. This is not to suggest, however, that there is not a wide range of other types attracted to librarianship.

In 2000, an analysis of 66 permanent staff at the UWL confirmed the modal type for the Library to be ISTJ. The next most popular type was ISFJ. Other types in the sample included at least one representative of all but one of the 16 types identified by Myers and Briggs.

Without wishing to give undue emphasis to the predictive uses of type, the MBTI has proved a valuable tool for both individual and organisational awareness. It has been used with particular benefit in team building, communication and change efforts. Myers *et al.* (1998), and other sources, provide useful strategies for managing change and for raising organisational and individual awareness of the value of difference.

The most valuable lesson to be learned from type theory is that a balance of the different types is desirable, especially in problem solving and decision making. "The theory of psychological type suggests that the best decisions include using all the perspectives identified by the MBTI functions (Sensing, Intuition, Thinking, and Feeling) and experience with groups in organizations confirms this." (Myers *et al.*, 1998, p. 339).

If, as Myers suggests, "...rapid adaptability to change, flexibility and the like, are not typically comfortable and natural parts of the ISTJ personality," (Myers *et al.*, 1998, p. 67), then this awareness is usefully translated into relevant communication and training styles. Additionally, if assurance is given by the organisation's leaders that the preferences of all staff are equally valuable and valued in the workplace, then the strengths of the ISTJ can contribute not only to change effectiveness but to the full range of processes and projects. ISTJs are characterised by Myers *et al.* as practical, sensible, systematic and realistic with a logical, fact-based approach to decision making.

It is probable that at least some members of library staff are drawn to the profession by a desire to assist and provide service to others. It is interesting that the modal type for the UWL Lending Services Team is ESTJ. With the substitution of a preference for extraversion over introversion, ESTJs enjoy interacting and working with others while still valuing facts, logic and pragmatism.

From experience, however, it is clear that a service ethic cannot be assumed to be present and must be defined, communicated and instilled in staff in various ways, including measurement, evaluation, training and development.

VI. Challenges and Insights

A. Equipping Staff to Measure

The only categories of the Framework, which consumed more time on the "quality journey" than client satisfaction and the associated concept of client relationship management, were the data analysis and quality process categories. The skill set least in evidence amongst Library staff included the ability to analyse and graphically present data and other information and the understanding of statistical variation.

> A profession that sees itself as "doing good" is less concerned with outcomes and impacts, since it sees its activities as inherently positive. Assessment activities also require a certain skill set, which has not been readily available to the profession. (Lakos, 2001, p. 313)

Education for librarianship has not by and large equipped library staff with the requisite skills to conduct measurement and evaluation. Statistical and data analyses are neither commonly taught, nor are decision making, problem solving or financial analysis and reporting. To succeed in implementing the Framework, to demonstrate improvement and to measure and benchmark performance, it was essential for staff to acquire the necessary skills.

Training programs, first led by an external consultant and then developed internally, introduced staff to the so-called "quality tools" and the basic concepts associated with measurement and evaluation. This same skill set was vital in the development of performance indicators and measures. An organisation-wide approach was mandated as leaders and staff alike lacked the requisite knowledge. Team members, their leaders and the university librarian all participated in workshops to firstly, map core processes and secondly, to develop agreed indicators and measures for all core functions and processes.

Staff members who had received external training developed workshops, tailoring examples to be more accessible and applicable to a library setting. A lengthy, often arduous process resulted in an initial set of indicators which have been reviewed, revised and refined over time. On reflection, it was probably a benefit that all learned together and that staff could see that their leaders were genuinely grappling with the same new ideas as they were.

B. Development of KPIs

Since it was intended that team members would be responsible for conducting, analysing and reporting their team's performance, ownership of the indicators and the measurement process was critical. In many organisations a quality manager or similar position is responsible for the analysis and reporting process. A position of Quality Coordinator was established at UWL in 1996, however, all teams continued to be responsible for their own measurement and reporting, seeking advice from the Quality Coordinator only when needed.

Measures to support indicators such as "document delivery frequency" could be relatively easily constructed to include "fill rates" and "turnaround times." More demanding was the development of indicators in the Framework's category of "Business Results." Benefits or value to the Library's community and stakeholders could not readily be expressed in financial terms. The question of value measurement for information and educational "products" is discussed by McGregor (2000) and concludes with the acceptance of Broadbent's (1991) position that: "When the real impact of an information system cannot be measured, the perceived value may have to be accepted as a proxy. The perceived value approach is based on the subjective evaluation by users and presumes that users can recognise the benefits derived from an information service..." (Broadbent and Lofgren, 1991, p. 98).

Ultimately, UWL adopted one Key Performance Indicator (KPI): *Client and Stakeholder Satisfaction*. All other indicators and measures provide data and other information to support this KPI. Evaluation of performance, however, does not rely on client perception of satisfaction alone. Business-oriented performance indicators were developed to measure, for example, supplier performance, budget utilisation and facilities use.

It is recognised that, despite the development of many robust and useful indicators, some measures, such as surveys and feedback incidents, are essentially measures of perception. Conscious that the development, administration and interpretation of indicators can be time-consuming and must be justified in an environment of resource constraints, research continues on identifying indicators which are objective and able to withstand academic scrutiny. As flexibility and change agility may be the key determinants of future organisational sustainability, effort was directed also to developing a sustainability model and the beginnings of indicators to measure these concepts, see Fig. 5.

C. Managing the Change Process

Using the ABEF as a reference point reinforced that performance measurement is only one element in an integrated structure which includes establishing future direction or vision, developing goals and strategies to realise the vision, ensuring implementation through action plans, determining how success will be measured and then feeding the results of evaluation back into the planning and improvement cycle.

Knowledge and experience of quality management was identified as an essential competency for all middle and senior managerial roles. Quality awareness is a core training requirement for all staff. Potential recruits and existing staff are thus aware of the continuing importance of this skills set.

At UWL, the Library executives were familiar with change management theory and had some experience in its practical application. Preparation for the introduction of the ABEF, therefore, followed established principles. This approach was vindicated by the widespread acceptance of the "Quality Service Excellence" program amongst staff. Their responses, in the main, ranged from compliance to enthusiasm. Some took more time than others to participate in developing team plans and performance indicators and to actively embrace the quality program.

A brief discussion of "change" principles which were found to be particularly effective follows. Of primary importance is that leaders should have realistic expectations of the length of time needed to implement and integrate an organisation-wide change, involving, as the UWL quality program did, cultural change, individual and team development and significant learning of new skills and knowledge.

Before announcing a new direction, it is important to have the commitment of the senior management team and ensure that they understand and can explain the philosophy and purposes of the program in a way that takes into account the different learning and communication styles of staff members. Inevitably, some cynicism with the new, business-oriented terminology will be apparent. In these cases, it is worth taking the time to explain terms and concept with examples of library application. Identifying gatekeepers or change agents, those staff who are change oriented and who are influential with other staff, is valuable in disseminating the message in more informal contexts.

The importance of systematic communication cannot be overemphasised. Effective communication depends on the development of a variety of mechanisms, processes and actions, enabling the dissemination of consistent messages on a planned and regular basis. Successful internal communication is perhaps the most often criticised aspect of organisational management and the most difficult goal to achieve. This was the case at UWL and internal communication has been the subject of a quality improvement team, as well as regular review and improvement actions.

D. Rewards and Recognition

Leaders and gatekeepers alike should be encouraged to model the values and behaviours associated with the change effort. Recognising and rewarding active participants in the process provides visible symbols of the espoused values. Incentives to change, in the form of awards and other recognition, are motivating for some staff. The question of performance-related rewards in libraries is yet another challenge as rewards which are common in business organisations, such as bonuses, promotion and fringe benefits, are not readily available to Library managers. It is possible, however, to try to understand what it is that rewards people and how this differs amongst individuals.

Leaders tend to make assumptions about what is rewarding, as was the case at UWL. After an initial round of awards for excellent service and exceptional performance, a brief staff survey was administered and the results were used to tailor rewards to meet the most commonly expressed preferences. Publication of a rewards and recognition policy and leaflet helped ensure that the process and the criteria for each form of recognition were transparent. A rewards scheme assisted in engendering a competitive spirit, particularly amongst teams which, in turn, encouraged higher performance and increasing comfort with the more competitive environment in higher education libraries, discussed earlier in this chapter.

A further insight from the UWL experience was that for many staff, the less tangible benefits of the application of quality principles were significant. The opportunity to develop new skills, for instance, in data collection and analysis and process and project management, provided a new level of interest for many staff engaged in routine jobs and, for some, career progression opportunities.

Concepts associated with the ABEF such as "empowerment" were motivating also for many staff. "Empowerment" is one of those management terms which is not readily acceptable to all. Open discussion of the concept with all staff to explore and agree on the intended meaning for use in the UWL context contributed to staff acceptance. The development of a model, see Fig. 2, to illustrate the shared meaning which had been developed also helped. Many staff, in fact, enthusiastically embraced the concept and the opportunities it presented for taking initiative, making decisions and planning their work schedules. Success of the empowerment strategy was largely due to the knowledge management structures and communication strategies which supported it.

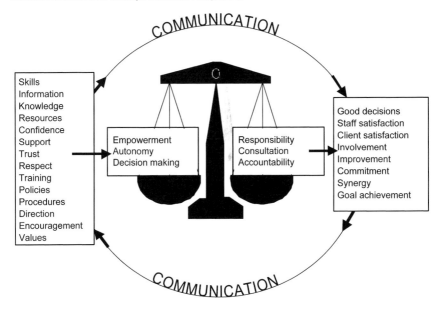

Fig. 2 Empowerment—finding the balance.

Knowledge and learning were incorporated in the values and "Ideal Culture," discussed below, and represented in the vision, mission and goals. The importance of continuous learning underpinned most human resource management strategies. The inclusion of core competencies in performance management processes was aimed at linking evaluation and learning in a process intended to be developmental for both staff member and team leader.

Ultimately, becoming the first Library to win an Australian Business Excellence Award provided recognition throughout the University and the profession and this was rewarding and confidence building for all Library staff.

Of the personal attributes useful to those planning to lead change of such magnitude, persistence, focus on the vision or planned outcomes and sense of humour would rank amongst the most important. Persistence in achieving the envisaged change is essential for many reasons, not least to counter the possibility of "quality" being labelled as the latest management "fad".

Availability of resources, such as staff time and dollars, is an obvious though sometimes overlooked planning prerequisite. The required resources are not great. Funding allocated to the implementation of a quality framework at UWL was insufficient to make a difference in terms of traditional competitive indicators such as collection size. It was possible, however, through following the precepts of the ABEF to achieve many challenging goals, to introduce new services and process improvements and to achieve a culture of assessment, all of which combined to achieve a competitive edge for UWL.

VII. Organisational Culture

> The concept of culture relates to the ideas and assumptions which are developed by people in any social group and which have a major impact on group behaviour and judgments. (Dawson and Palmer, 1995, p. 167)

In a climate of constant change, growing competition and expanded student expectations of services, organisational flexibility is paramount. Traditionally, libraries are highly structured, regulated and hierarchical with defined departmental boundaries. Perhaps library organisation charts reflect the rule-based structure of cataloguing and classification systems and the adherence to codes essential to the effective retrieval of information.

One of the early initiatives introduced at Wollongong to reduce the emphasis on hierarchy was to design a flatter, more flexible structure with teams as the primary structural unit. This was not simply a matter of regrouping or relabelling. The process, referred to as "team building" included extensive preparation and training, in recognition of the reluctance of some staff, accustomed to working more or less alone, to become team players. In many cases, teams were extended to include previously separate functions. Interaction and cross-fertilisation were further encouraged by the formation of quality improvement teams to address perceived performance gaps. Standing committees to manage staff training and development, client services and quality assurance were also formed. All teams included representatives from different groups and levels of staff.

An empowerment model underpinned team building, see Fig. 2. Teams were able to establish their own objectives, in accordance with broad Library goals, could modify team processes and were able to solve problems as a team, using quality processes and methods. As well as establishing the conditions for goal ownership and achievement, a more favourable climate for flexibility and process improvement was put in place. The concepts of "teams" and "teamwork" became building blocks for the organisational culture, characterised by staff commitment to assessment, goal achievement and flexibility, cultural hallmarks which the Library executive aspired to develop.

"TQM is the first managerial movement that has specifically considered culture and the values that develop in an organisation" (Dawson and Palmer, 1995, p. 55). The culture envisaged in TQM theory is one that supports flexibility, continuous change and commitment to organisational goals. Roadblocks presented by a hierarchical structure, rigid top-down management and minimal input by employees into decision making, were familiar to those involved in early change efforts at UWL. The Library executive strongly believed that cultural alteration and improvement were both necessary and possible; and that they must accept responsibility for leading the change and for modelling the attributes and behaviours conducive to the envisaged culture.

By 1994, progress had been made in addressing the constraints of a culture characterised by conservatism, hierarchy and resistance to change. The key initiatives of the 1980s and the early 1990s had been strategic planning, team building, staff development and performance management. Most staff responded positively to the first three although opposition from some staff to a team-based structure was surprisingly persistent. The introduction of performance appraisals for all staff, although carefully researched and sensitively managed, caused a greater level of apprehension and required a longer lead-time than team building. A robust performance management system was critical to the achievement of assessment dimensions in the envisioned culture as it introduced the first tangible elements of evaluation and improvement. Staff gradually started to accept the notion of accountability for goal achievement as well as recognising the benefits of relevant skill development.

Application of the principles embedded in the ABEF enabled changes to be integrated into the "normal" workflows of the Library. Longevity was envisaged for the ABEF as UWL's management model and quality assurance was not to be construed as distinct from other processes. In essence, the aim was to drive a cultural change which would achieve a sense of overall responsibility for organisational performance, previously considered to be mainly within the province of the library executive.

The articulation of shared values formed the basis for identifying those shared beliefs and norms which underpin organisational culture. In developing the initial set of values, most staff agreed that valuing clients, teamwork and continuous improvement, for example, should be included. A second iteration of the values changed the focus to "satisfied clients," "partnerships" and "open communication." In the latest review in 2002, a number of the previous values were felt to be so well integrated that they could be omitted from the published values statement. Instead, "Satisfied Clients" was replaced with "People First" to embrace both external and internal (staff) clients; "Knowledge and Learning" was replaced with "Sharing Knowledge and Learning" to capture commitment to organisational learning. One impetus for reviewing the values was to examine their congruence with the notion of an "Ideal Culture".

Understanding and internalisation of the "Ideal Culture" was enhanced by defining each of the values with behavioural examples, supplementing the concepts with personal attributes, such as "approachable," "self aware" and "flexible" and with performance attributes appropriate to each staff group. Both sets of attributes were incorporated into Position and Person Profiles, previously known as position or job descriptions. The performance management process was revised to include evaluation against the attributes. The "Ideal Culture" is described in UWL brochures as: "the working environment to which we aspire, in which every staff member strives to uphold the Values, is actively developing the desired personal Attributes and is building their knowledge and skills to achieve relevant Performance Attributes." Implicit in the last part of the definition is the valuing of assessment at individual, team and organisational levels.

As described in McGregor (2003, p. 8), "All staff received extensive in-house training in quality tools and techniques and participated in self-assessment exercises and numerous surveys. Most importantly, all staff contributed to the development and review of vision, mission, goals, values, performance indicators and measures. Each team was, and remains, responsible for administering and reporting its own measures. Although this was challenging in many instances, the outcome of the process was reinforcement of a long-standing goal: to develop a culture of commitment and assessment."

The importance of a culture of assessment in bringing about change in libraries is discussed by Lakos (2001). Lakos defines a culture of assessment as: "...an organizational environment in which decisions are based on facts, research and analysis, and where services are planned and delivered in ways that maximise positive outcomes and impacts for customers and stakeholders" (Lakos, 2001, p. 313). He goes on to articulate a number of conditions which should be present for a culture of assessment to develop. These include focus on customers' needs, the inclusion of performance measures in plans, the commitment of leaders to assessment and external focus.

Enabling staff to understand external forces affecting the Library and its plans or, in other words, to "see the big picture," was a key strategy in the cultural change which accompanied introduction of the ABEF. The strategy included explanation of environmental influences at meetings, sometimes by expert guest speakers, attendance at conferences and in-house seminars, staff involvement in SWOT analysis and strategic planning and dissemination of information via multi-layered communication mechanisms. Of primary importance was the articulation of knowledge management policies and processes to support the sharing of knowledge, information and experience.

These elements all formed part of integrated strategic planning and environmental awareness processes. Relating general trends to the impact on specific library tasks and services, as well as emphasising the importance of every team's involvement in adapting their functions to relevant external changes, was revelatory for some staff. Of benefit also was that staff were better equipped to participate in faculty planning and university working groups, thus making progress towards one of the Library's goals of increased involvement in the University's planning processes.

Broad knowledge of the higher education environment and awareness of external forces were of the greatest importance in the development of all goals, strategies and performance indicators. As Cullen (1999) explains:

> Performance measurement is a highly political activity and must be seen as such, at the macro or micro level. We must look outwards to social and political expectations made of our institutions and ensure that they meet the needs and expectations of our significant client or stakeholder groups; we must use our planning and goal-setting activities in a meaningful way, incorporating appropriate measures, to demonstrate our response to this external environment, and our willingness to align our aspirations to broader corporate goals. But we must also look within and seek to promote an organizational culture which acknowledges the political nature of measurement. This means using performance measurement to:
>
> - indicate the library or information service's alignment with broader organizational goals;
> - demonstrate the integration of information services with the key activities of the organization, or of the community;
> - support the library's position as the organization's primary information manager and service provider. (Cullen, 1999, p. 25)

Given the importance of developing flexibility or adaptability in the organisational culture, it was necessary to go beyond continuous improvement and provide cultural support for risk taking and innovation. Policy and procedural support was accompanied by the introduction of an award for innovation. The organisational value, "initiative," is defined in terms of supporting risk taking, learning from mistakes and looking for solutions and innovative ideas. "Flexibility" is an agreed personal attribute and both personal attributes and values are components of the "Ideal Culture" discussed above. Flexibility or adaptability is critical as discussed by Heskett *et al.* (1997): "…the single most important indicator of adaptability was the adherence by management to a clear set of core values stressing the importance of delivering results to various constituencies, especially customers and employees…" (Heskett *et al.*, 1997, p. 250).

Heskett describes a project encompassing 200 firms in 19 industries to examine the relationship of performance and culture and found that: "The clear differentiator between high and low performing firms, all with strong cultures, was the ability of each firm to adapt to changing environments…" (Heskett *et al.*, 1997, p. 250). It was observed that organisations which "install devices for maintaining adaptability not only greatly improve their chances of sustaining high performance over time, they increase their chances of achieving successful transitions from one leader to another" (Heskett et al., 1997, p. 250). The "device" chosen by many firms was continuous quality improvement "forcing an organization to compare itself with the best performers and generally become less insular in its thinking" (Heskett et al., 1997, p. 250).

VIII. Benchmarking

Within the Australian Business Excellence Framework (2003, p. 44), benchmarking is defined as: "a method of comparing and measuring processes and outcomes with those of recognised leaders, with the intent of improving performance." At UWL benchmarking has been used for process improvement and for comparing performance with organisations regarded as "best practice," as assessed by the Australian Business Excellence Awards evaluators. Comparing or benchmarking performance may often lead to improved process efficiency, however, the primary benefit is the identification and harnessing of good ideas and applicable strategies, as well as stimulating critical thinking about all of the library's activities.

At the beginning of the "quality journey," benchmarking was an unfamiliar concept. As with all elements of the ABEF, it was important for the Library to determine its approach to benchmarking, in other words, the intentions and desired outcomes which would underpin benchmarking activities. Secondly, implementation or deployment was planned; who would be involved, what training would be provided and how benchmarking partners would be identified. Thirdly, how the indicators and measures identified for benchmarking success would be monitored and evaluated was determined. Lastly, following evaluation and analysis of benchmarking projects, how results would be incorporated into future planning was established.

The process outlined in the previous paragraph illustrates the Approach, Deployment, Results, Improvement cycle which the ABEF has developed to guide the design of systems and processes, see Table III. Known as ADRI, the cycle is also used as an assessment matrix for award applicants in preparing their submission for an Australian Business Excellence Award and in the subsequent evaluation visit.

UWL has engaged in process benchmarking with other university libraries and in both process and organisational level benchmarking with organisations outside the library sector. While there is value in benchmarking with peer organisations, it may be to the detriment of libraries if they exclude other organisations, which are recognised leaders outside the sector, as potential network partners simply because they are perceived as being "different".

Table III
ADRI (ABEF, 2003, p. 30)

Approach	*Thinking and planning.* Identifies the organisation's intent, the thinking and planning it undertakes to design the strategies, processes and infrastructure to achieve the intent, including the design of performance indicators to track progress
Deployment	*Implementing and doing.* Describes how strategies, structures and processes have been put into practice
Results	*Monitoring and evaluating.* Demonstrates how measures or achievement associated with the Approach are monitored and examines trends in performance
Improvement	*Learning and adapting.* Examines what has been learned and how this learning is used to improve the approach and deployment

Organisations of all types, including higher education institutions, are grappling with issues such as: demonstrating value, managing scarce resources, managing client relationships and meeting changing client and stakeholder needs and expectations. Competition, market differentiation, partner and supplier relationships and future viability are as applicable in higher education today as they are in global corporations. While organisations outside the library and information sector may have different goals and objectives, what is being sought through benchmarking is creative, innovative solutions to common issues and problems.

The issue of understanding, valuing and managing organisational differences highlights what benchmarking is not, that is, adopting another's practices verbatim as a solution to a problem. This holds true for good or best practices observed in another library. Benchmarking can provide the necessary catalyst for adjusting, adapting or modifying practices to best suit the specific needs of the library, while adoption of any changes should take into account the mission, goals and environmental constraints which are unique to each organisation.

Benchmarking with other libraries has been an important activity to focus attention on continuous improvement. Processes that have been scrutinised by UWL include: document delivery, acquisition of resources, cataloguing, loan returns and shelving practices. Working with high performing libraries in these areas has highlighted opportunities to re-engineer processes, consider new or different technologies, or simply eliminate steps in processes which no longer add value to the final product or service.

Partnering with recognised leaders outside the library and information sector, however, has provided the opportunity to observe continuous improvement and innovation initiatives beyond established organisational paradigms, providing the necessary stimulus to actively challenge and question the efficacy of existing practices. Participation in the Australian Business Excellence Awards enabled access to partners who had already been recognised as "best practice" organisations against established criteria and principles of excellence. Despite the diversity of organisations within the benchmarking networks such as legal, pharmaceutical, telecommunications and health care organisations, all share common issues and processes. Examples of processes examined within these benchmarking networks included: internal communications, management of change and human resource management.

Of particular interest was a project conducted jointly with the Wollongong City Council which is responsible for local government in Wollongong. The Council was recognised with an Australian Business Excellence Award in 1998. The purpose of the project was to examine customer perception of value when a customer is engaged in a service transaction and to determine whether core values were common across a variety of service scenarios.

A number of issues must be addressed within the ABEF's "customer perception of value" item. These include: how organisations measure whether or not customers believe they have received fair value and how organisations communicate customer perception of value in order to help staff at all levels contribute to achieving customer satisfaction goals.

Joint research activities included focus groups and surveys of a broad cross-section of the community including: university students, parents with young families, single income earners, disabled citizens and aged pensioners. Participants in the study identified a series of common value attributes that were considered important by each segment across the four service scenarios. The positive relationships and commonality found amongst survey respondents provided valuable input into the development and delivery of customer service skills training and the management of customer service relationships for both the Library and the City Council.

Internally, benchmarking delivered a number of benefits for UWL. These included improved understanding of internal systems and business practices; establishment of key success factors and measures of productivity; new ideas leading to either continuous improvement or breakthrough change and improvement in understanding and meeting the needs of clients. Sharing and discussing the results of benchmarking and evaluation was salutary for many library staff who thought that processes were already as efficient as possible.

Externally, benchmarking and networking with non-library organisations has expanded awareness of libraries, their current roles and the challenges they face which, it would be fair to say, remain relatively unknown outside the profession. As an Award-winning organisation, the Library has welcomed visitors and prepared presentations for those taking part in "Business Study Tours" organised by the Awards administrators. Visits and presentations provided opportunities to showcase library achievements, resulting in increased respect and recognition from those participating in the tours and their parent companies.

IX. Success and Sustainability

A. Indicators of Success

As discussed in Section IV, integral to the decision to commit to the ABEF was the intention to evaluate the effectiveness or otherwise of all the various change efforts by applying for an Australian Business Excellence Award. The process involved preparation of a 50-page submission outlining progress against all categories of the Framework. A team of accredited evaluators conducted an audit against the submission, followed by an on-site evaluation. An extensive feedback report was provided to applicants, regardless of whether the organisation was to be recognised with an award. This external feedback was invaluable to UWL in both identifying improvement opportunities and in recognising strengths.

Although the Business Excellence Framework was found to be largely applicable to non-profit organisations, some aspects were more challenging than others. For example, Category 7: "Business Results" consists of two main sections or items "indicators of success" and "indicators of sustainability." The expectation in this category is that overall organisational performance, both in the present and as predicted in future, will be demonstrated.

Broad indicators of financial performance available to corporations, such as profit margin and return to shareholders, are not applicable to libraries. However, financial performance can be measured in terms of effective and efficient budget management, including strategic fund allocation, as well as a range of other broad quantitative and qualitative indicators and measures. The following is an extract from UWL's award submission (University of Wollongong Library, 2000), which articulates the approach taken to measuring and demonstrating organisational performance followed by some examples of "indicators of success".

- The Library's KPI is *Client and Stakeholder Satisfaction*. All contributing indicators measure significant components of our business and can be aggregated to evaluate overall organizational performance. Satisfied and supportive clients are the hallmarks of our success. Satisfaction of

stakeholders with our strategy, management framework and recognition by those outside the University indicate success in the business dimensions of our overall performance.
- Overall performance is assessed using lead and lag indicators, depicted in our PIF. We determine the PIF's fitness for purpose by the reliability and relevance of data collected; ease and speed of extracting and extrapolating data and information and ability to measure stakeholder expectations.
- We also use the PIF as a diagnostic tool to predict future performance, e.g., indicators such as leadership effectiveness, budget utilisation, application of innovation and technologies and acquisition of skills and knowledge.
- We recognise that it is critical to improvement efforts and to our goal of leadership in the information industry, to benchmark against external standards whenever we can identify those relevant to our business.
- Teams monitor lead and lag indicators on a monthly basis, the Library executive reviews key data and indicators quarterly and formal reports are prepared on a half-yearly and annual basis. Performance is also reviewed and recorded in the Annual Report to University Council; this provides the opportunity for key stakeholders to comment on our progress.
- Goals and critical success factors are developed to meet stakeholder expectations and to make progress against strategic initiatives. Performance indicators are used to evaluate success and have been developed for all goals and critical success factors.

Some examples of results cited in the submission by client and stakeholder category include:

Stakeholder Group 1—University Executive

- Process improvement, through analysis of measurement data, benchmarking outcomes and feedback from external evaluation, resulted in an overall processing cost reduction of 22% since 1997.
- Salary costs have been effectively contained while providing the highest level of investment in information resources. Savings have been redeployed to technology investment and planning future services.
- Planning success is evidenced by over 80% of stated actions being achieved within the planning cycle each year.
- Leadership effectiveness is demonstrated by benchmarking against other university leaders.

Stakeholder Group 2—Clients

- The Academic Outreach program is founded on personalised contact and coaching to optimise new products and services.
- Client satisfaction with our services is high and we continue to set improvement targets.
- Client feedback illustrates our success in applying business excellence principles, e.g.: *I wish the rest of the University would have similar standards. Keep using the ABEF—it pays for you and your clients.* (Alexander Hausner, November 1999).
- Success is dependent on our staff to a great extent: on their commitment, readiness for change; excellent client skills and their initiative, see Fig. 3.
- Achieving the *Investors in People*[2] standard has benchmarked the Library against world-class organizations.

Stakeholder Group 4—Suppliers

- Working closely with suppliers has developed mutual understanding of requirements; raised their performance levels in meeting our needs for timely, accurate, cost-efficient supply; and developed innovative solutions, see Fig. 4.

Stakeholder Group 5—The Community

- Secondary school penetration rates have improved by 33% since inception in 1998, and student confidence levels in using Library services after the program have been maintained at 95–98%.

B. Indicators of Sustainability

The second item in Category 7, "indicators of sustainability," is concerned with the collection of information to predict likely future relevance and viability. Again, some of the business-related concepts such as market trends were not immediately accessible. They were, however, concepts which were considered long and hard leading to, for instance, the development of

---

[2] For a description of the "Investors in People" Standard and its application in the University of Wollongong Library, see the following publications:Denny, Lorraine (2000). University of Wollongong Library—Investors in People. *AIMA News*, 15 June, p. 1, 4. Denny, Lorraine (2000). What "Investors in People" have done for us. *NATA Certification Services International*. Jantti, Margie (2000). Investing in people. *Momentum, The Quality Magazine*, 1. Or the following Website: http://www.ncsi.com.au/.

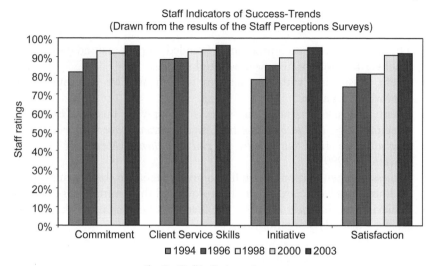

Fig. 3 Staff indicators of success.

a marketing plan which documented the various client groupings or segments and the services available for each. Additionally, the plan facilitated identification of groups needing additional specialised services. This in turn enabled progress towards achievement of one element of UWL's vision, that is, to provide exceptional service, customised to meet individual needs.

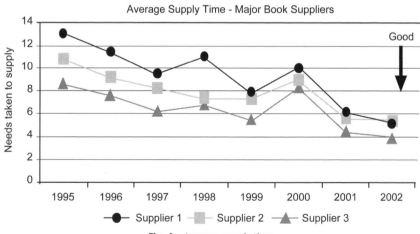

Fig. 4 Average supply time.

The following extract (University of Wollongong Library, 2000) articulates the approach taken to demonstrating sustainability.

- Best practice studies in the higher education sector indicate that dynamism is as important as past achievements and is probably a better guide to future performance.
- We have positioned ourselves to be an indispensable partner in the University's business. The Library is widely regarded as the heart of the University, as a model of excellence in service and as possessing a "different" culture.
- Acutely aware of the volatile and increasingly competitive environment in which we operate, and of the many risks involved in choosing one strategy over another, we intend to continue collection and analysis of all factors which impinge on our business so that choice and positioning is based on the best possible information.
- We have identified the factors which will influence our ongoing success and incorporated these in our planning processes.
- We keep key stakeholders informed of how we add value for mutual advantage and of strategies for managing threats and opportunities, e.g., providing expert input into planning campus-wide strategies, such as flexible delivery, internet access, research infrastructure, generic skills, human resource policies.
- We form worldwide alliances to influence publishers, suppliers, other information providers and potential competitors, by working with them to determine roles, future pricing, service models and by aligning ourselves with influential partners. We work through consortia to negotiate the best possible coverage and pricing of resources.
- We invest in our staff, in skills and knowledge acquisition and in leadership development, in relevant and carefully tested technology and innovation in all aspects of the business, as these are critical to our sustainability.

A model was developed to illustrate the approach and how it would be measured, see Fig. 5.

Addressing and implementing the precepts of Category 7 provided extensive learning opportunities at individual and organisational level. To realise that sustainability into the future cannot be assumed, even for "public good" organisations such as libraries was valuable in itself. Library staff became increasingly aware of the competition posed by Internet search engines, commercial online learning providers and some information technology staff who believed that mastery of systems and search software is all that is needed for successful information retrieval in the online era. Another benefit was found in analysing and demonstrating the tangible and intangible value the Library is able to deliver to its clients and stakeholders.

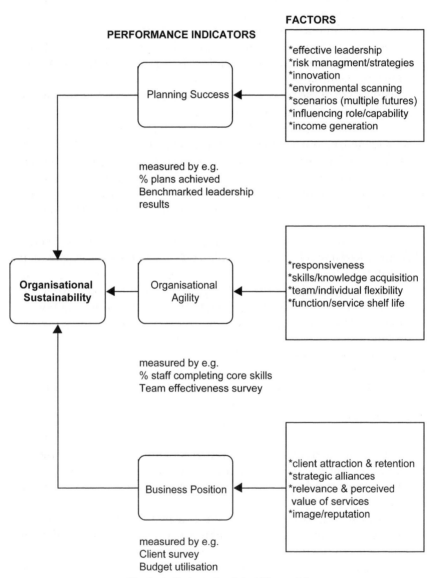

Fig. 5   Indicators of sustainability model.

X. Conclusion

As demonstrated throughout its submission for the Australian Business Excellence Award, the UWL had measured performance internally and externally, both within and outside its industry through benchmarking and evaluation against national and international standards.

In its pursuit of excellence UWL adopted a quality management framework, as a matter of choice not, as is often the case, to comply with government or parent body fiat. As outlined by Crosling (2003, p. 43), UWL adopted "...an *intrinsically motivated* approach—one driven by shared values and aspirations" as opposed to "an *extrinsically motivated* approach—one driven by imposed rules and regulations...".

The many benefits accrued during the UWL's quality journey have been presented in other papers by McGregor (1997, 2000, 2001, 2003). Some of these benefits and insights have been delineated in this chapter. Perhaps the defining benefit is that the decision to embark on a quality program through adoption of the ABEF has been validated over time. It is now almost 10 years since the commencement of the journey. The principles and philosophy of the ABEF have been integrated into the work of all teams; "quality" has become part of the way of "doing things." This has translated into the recognition for service excellence and the competitive positioning of the Library envisaged at the outset. There is confidence in the ability of the Library's staff, working together as well as in partnership with clients and stakeholders, to identify solutions to whatever challenges both the Library and its parent organisation may face in the future.

Application of the principles and the assessment dimensions: Approach, Deployment, Results and Improvement has encouraged a collective focus on the core competencies of the Library; that is, what the Library must do well for the achievement of its vision, mission and goals. The ABEF, while providing insight into good and best practices, does not prescribe what organisations *should* do to achieve competitive advantage. Each organisation must design and execute the systems and structures which best fit its mission, its operating environment, stage of development and its range of situational variables to achieve the best possible strategic advantage.

As discussed by Barney (1995), the challenge for any organisation is the development of a management configuration that simultaneously exploits and develops the core competences and efficiency practices in a manner that is not easily replicated by its competitors. Despite the ease of imitating processes and practices, organisations that are successful in strategy selection and execution are often leaders in developing complex, sophisticated and often intangible competences and resources, such as organisational culture, features not prone to easy duplication.

The anticipated indicators of success will be tested by the critical audience of the intended strategy; whether the users of the system can be convinced that it will deliver best value and services to its customers and other key stakeholders (Johnson and Scholes, 2002).

References

Audit Manual (2002). Australian Universities Quality Agency. Unpublished, http://www.auqa.edu.au/qualityaudit/auditmanual_v1/index.shtml.
Australian Business Excellence Framework (2003). Standards Australia International, Sydney.
Barney, J. B. (1995). Looking inside for competitive advantage. *Academy of Management* 9, 49–61.
Broadbent, M., and Lofgren, H. (1991). *Priorities Performance and Benefits: An Exploratory Study of Library and Information Units*. Centre for International Research on Communication and Information Technologies & Australian Council of Libraries and Information Services, Melbourne.
Carnegie, G. D., and West, B. P. (2003). Placing monetary values on public repositories of knowledge: why, when, how and so what? Paper presented at the *Council of Australian University Librarian Seminar*, Cairns, Australia.
Crosling, R. (2003). *Using the Australian Business Excellence Framework to Achieve Organizational Excellence*. Standards Australia International, Sydney.
Cullen, R. (1999). Does performance measurement improve organizational effectiveness? A postmodern analysis. *Performance Measurement and Metrics* 1, 9–30.
Dawson, P., and Palmer, G. (1995). *Quality Management: The Theory and Practice of Implementing Change*. Longman, Melbourne.
Groenewegen, H., and Lim, E. (1995). TQM and quality assurance at Monash University Library. *AARL Australian Academic and Research Libraries* 26, 6–16.
Hernon, P., and Whitman, J. R. (2001). *Delivering Satisfaction and Service Quality: A Customer-Based Approach for Libraries*. American Library Association, Chicago, IL.
Hernon, P., Nitecki, D., and Altman, E. (1999). Service quality and customer satisfaction: an assessment and future directions. *Journal of Academic Librarianship* 25, 9–17.
Heskett, J. L., Sasser, W. E., and Schlesinger, L. A. (1997). *The Service Profit Chain*. Free Press, New York.
Johnson, G., and Scholes, K. (2002). *Exploring Corporate Strategy: Text and Cases*. 6th ed., Financial Times/Prentice Hall, New York.
Lakos, A. (2001). Culture of assessment as a catalyst for organizational culture in libraries. In *Meaningful Measures for Emerging Realities: Proceedings of the 4th Northumbria International Conference on performance Measurement in Libraries and Information Services* (J. Stein, M. Kyrillidou, and D. Davis, eds.) pp. 311–319. Association of Research Libraries, Washington, DC.
McGregor, F. (1991). Managing organizational change: an action research case study. In *Proceedings of the 1st World Congress on Action Research and Process Management*. Acorn Publications, Brisbane.

McGregor, F. (1997). Quality assessment—combating complacency. *Australian Library Journal* **46**, 82–92.

McGregor, F. (2000). Performance measures, benchmarking and value. In *ALIA 2000 Capitalising on Knowledge the Information Profession in the 21st Century: Proceedings*. ALIA, Canberra, [online] Available URL: http://conferences.alia.org.au/alia2000/proceedings/felicity.mcgregor.html [accessed 26/2/04].

McGregor, F. (2001). Inside, outside and upside down. Paper presented at the *International Business Excellence Summit*, 25–27 March. Sydney Convention and Exhibition Centre, Darling Harbour.

McGregor, F. (2003). Benchmarking with the best. Paper presented at the *5th Northumbria International Conference on Performance Measurement in Libraries and Information Services*, Durham, England.

Myers, I. B., McCaulley, M. H., Querk, N. L., and Hammer, A. L. (1998). *MBTI Manual: A Guide to the Development and Use of the Myers–Briggs Type Indicator*. CPP, Inc., Palo Alto, CA.

Quinn, B. (1997). Adapting service quality concepts to academic libraries. *Journal of Academic Librarianship* **23**, 359–369.

Schwartzman, R. (1995). Are students customers? The metaphoric mismatch between management and education. *Education* **116**, 215–220.

Shaughnessy, T. W. (1995). Total quality management: its application in North American research libraries. *AARL Australian Academic and Research Libraries* **26**, 1–5.

Sirkin, A. F. (1993). Customer service: another side of TQM. *Journal of Library Administration* **18**, 71–83.

University of Wollongong Library (2000). *Submission Award Level*. Unpublished.

The Civic Library: A Model for 21st Century Participation

Diantha Schull
Americans for Libraries Council, New York, NY, USA

> Librarians must become active not passive agents of the democratic process.
> (Archibald MacLeish, American Library Association Conference, 1940)

I. Introduction

In recent years there has been growing discussion in the library community regarding the civic role of the public library. The discussion is rooted in a deep-seated professional commitment to the value of the public library as an institution of democratic society. As a recent president of the American Library Association, Nancy Kranich, wrote in 2001, "Libraries serve the most fundamental ideals of our society as uniquely democratic institutions. As far back as the nineteenth century, libraries were hailed as institutions that schooled citizens in the conduct of democratic life." (p. vi).

Despite the vigor of this discussion, it has proceeded mainly in the academic and policy arenas, with minimal impact on day-to-day librarianship, library training, or library advocacy. Moreover, despite the implications of this discussion for the future of the profession, there has been relatively little examination of the practical strategies for developing the "civic library" or the "civic librarian."

This essay examines civic librarianship from a practical perspective. It focuses on examples of programs and other institutional initiatives in libraries across the country that offer approaches for civically and socially oriented services; it uses these examples as the basis for defining a service model, the Civic Library, and it examines the barriers and opportunities for operationalizing the Civic Library.

II. Context

In 1852, the trustees of the Boston Public Library articulated their goals for the new institution:

> The largest number of persons should be induced to read and understand questions going down to the very foundations of social order, which are constantly presenting themselves, and which we, as a people, are constantly required to decide, and do decide, either ignorantly or wisely. (Trustees, 1852)

One hundred and fifty-one years later, in 2003, Herb Elish, Director of the Carnegie Library of Pittsburgh, described his rationale for hosting the "Pittsburgh's Citizen's Deliberation," a deliberative opinion poll of over 80 people.

> Most people know Carnegie Library of Pittsburgh as a place to get books and research all kinds of subjects, but many may not realize the dynamic and active role the library plays in supporting a democratic and informed society...we look forward to offering the people of western Pennsylvania a center for the exchange of ideas on critical national and local topics. (Elish, 2003)

Despite the century-plus that separates them, these remarks by the first trustees of the Boston Public Library and the current director of the Carnegie Library of Pittsburgh clearly express an underlying commitment to the civic purposes of the public library. The continuity of this commitment over time can be traced through key texts in the library literature. In 1924, the Carnegie Corporation's study, *The American Public Library and the Diffusion of Knowledge*, recommended that the public library should serve as the "intelligence service" for its community, "not only for 'polite' literature, but for every commercial and vocational field of information that it may prove practicable to enter." (Molz and Dain, 1999, pp. 14–15). In 1947, Sidney Ditzion published his influential monograph, *Arsenals of a Democratic Culture*, in which he linked libraries to education through their function as a "people's university," through which "a wholesome capable citizenry would be fully schooled in the conduct of a democratic life." (Kranich, 2001, p. 30). Political scientist Robert Leigh, reporting in 1952 on the *Public Library Inquiry in the United States*, a study sponsored by the Carnegie Corporation, included the statement: "Public libraries continue to be of enduring importance to the maintenance of our free democratic society. There is no comparable institution in American life." (Kranich, 2001, p. 31).

On the international stage, the value of the public library as an instrument of democratic development has been affirmed repeatedly. The UNESCO public library manifesto states:

Freedom, prosperity and development of society and of individuals are fundamental human values. They will only be attained through the ability of well informed citizens to exercise their democratic rights and to play an active role in society. Constructive participation and the development of democracy depend on satisfactory education as well as on free and unlimited access to knowledge, thought, culture and information.

The persistence of a professional commitment to the civic role of the library is reflected in library mission statements, including these three, which can stand for hundreds of others:

Pasadena (CA) Public Library: "The freedom to know is the foundation of our democracy. The mission of the Pasadena Public Library, a basic municipal service, is to be an information center for the Pasadena community in order to preserve and encourage the free expression of ideas essential to an informed citizenry."

Brown County (WI) Library: "The Brown County Library system provides all residents of Brown County access to information and ideas from throughout the world in support of lifelong education, cultural enrichment, responsible citizenship, leisure activities, and economic development. The Library also contributes to this storehouse of knowledge by maintaining information unique to the area and its residents."

Tacoma (WA) Public Library: "The mission of the Tacoma Public Library is to provide the highest quality library services to fulfill the informational, educational, recreational and cultural needs of the citizens in the dynamic and changing community of Tacoma, which is comprised of many ethnic and economic backgrounds, and further, to recognize changes that occur in society and to adapt these changes to the delivery of people-oriented library services."

Another measure of the profession's consensus on the value of the library as a civic institution is the delineation of potential roles from which librarians should select, as expressed in successive versions of the American Library Association's *Planning Process for Libraries*. The ALA National Plan of 1947 stated:

> The objectives of the public library are many and various, but in essence they are two—to promote enlightened citizenship and to enrich personal life. They have to do with the twin pillars of the American way, the democratic process of group life and the sanctity of the individual person. (McCabe, 2001, p. 32)

Recent editions of *Planning for Results* list "The Commons" as one of the primary roles that a public library might select as a focus: "COMMONS: A library that provides a COMMONS environment helps address the need of people to meet and interact with others in their community and to participate in public discourse about community issues." (Nelson, 1998, p. 67).

Today, there appears to be renewed interest in the library as a civic institution. Two successive presidents of the American Library Association—Sarah Long (1999–2000) and Nancy Kranich (2000–2001)—selected complementary themes for their terms of office. Long focused on "Libraries Build Community" and Kranich focused on "Libraries: Cornerstones of Democracy." Both contributed to Kranich's collection, *Libraries and Democracy, The Cornerstones of Liberty*, which offered a series of essays exploring how libraries promote democracy, from a variety of disciplinary and theoretical perspectives. According to Kranich:

> If a free society is to survive, it must ensure the preservation of its records and provide free and open access to this information to all its citizens. It must ensure that citizens have the resources to develop the information literacy skills necessary to participate in the democratic process. It must allow unfettered dialogue and guarantee freedom of expression. Libraries deepen the foundation of democracy in our communities. (Kranich, 2001, p. v)

The growing focus on the "civic library" within the library community is matched by new awareness on the part of professionals from other sectors. For example, the Project for Public Spaces (PPS), a national organization that promotes preservation and design of public markets, parks and squares, has focused attention on the library's key function as one of the primary public spaces. At a national conference on parks organized by PPS in 1999, I was asked to speak on the relationship between libraries and parks as essential public spaces. Fred Kent and Phil Myrick of PPS, in an article in *American Libraries*, have stated, "the library is not just a research center but a place for community. Libraries have the potential to become an anchor of community life, even an attraction. Today, many libraries are evolving into multi-dimensional public spaces...a community front porch." (Block, 2003).

The architectural community has recognized the library both as a symbol and as an instrument for civic and social life. Many architects working on renovating older libraries to include contemporary functions, or designing new libraries, carefully address the need to include public spaces that promote interaction, learning and community collaboration. Architect Carole Wedge, of Shepley, Bulfinch, Richardson, and Abbott, which specializes in library design, states: "There's a longing for spaces in which to come together and be inspired...something you don't get from a laptop in Starbucks." (Morris, 2002). Demas and Scherer (2002), in their recent article "Esprit de Place," discuss the importance of the library as a community space and cite the "gossip corner" at the Detroit Lakes, MN, Public Library as an example of a "space for local citizens to meet informally—much like the agora of ancient Greece." Moishe Safdie, internationally recognized for his design of the Vancouver Library as the focus of a multifunctional "Library Square," cares deeply about the relationship between libraries and the urban context. Safdie describes the library as "not only a library but a major meeting place for the city—another city center." (Malouf, 2003).

In political science, libraries are the subject of analysis by experts such as Robert Putnam, a leading theorist on trends in civic engagement. Robert T. Putnam and Lewis M. Feldstein, a collaborator in the study of civic culture, recently published *Better Together*, a compilation of stories about key institutions. Their chapter "Branch Libraries: The Heartbeat of the Community," focuses on the Near North Branch of the Chicago Public Library as an exemplar of "The New Third Place." "As our glimpses of the branches in Chicago show, the new neighborhood library functions as a kind of community center, a place where people get to know one another, where communities find themselves." (Putnam and Feldstein, 2003, p. 49). Putnam and Feldstein also discuss the unifying value of the library, its inclusive and tolerant stance toward all, as an aspect of its value for local communities. They consider this unifying role as "one of the Chicago Public Library's core missions: to reflect and serve the diversity of the city's residents while helping those residents discover the sympathies and interests that unite them." (p. 54).

III. Paradox

Despite evidence of a relatively consistent consensus within the library community regarding the civic value of the library and despite significant external attention to the civic value and civic functions of the library, there is little evidence that the library profession has attempted to institutionalize these functions. The profession has not matched its rhetoric about the library as a cornerstone of democracy by creating a pedagogical or practical model for the library's democracy functions. There are no "civic librarians," per se, as there are Young Adult Librarians, Health Reference Librarians, Business reference specialists, or Children's Librarians. Librarians and library directors have little practical guidance from the profession when they seek to enhance their library's position as a civic institution.

In fact, the profession offers little clarity as to what constitutes civic library services. Does a library fulfill its civic potential simply by opening its doors? Beyond offering free and open access to all, what must a library do to activate this potential? Does the creation of a welcoming space make a "civic library?" Who is the civic librarian? As a result, when librarians act to protect public access to controversial information, they are certainly acting with the public good in mind, but does this alone make a library into a civic library?

Library literature has few answers to such questions. It offers no benchmarks for assessing if a library is acting in a fully civic manner. Despite the inclusion of "The Commons" as a library "role" in the Public Library

Association's recommended planning process, it has no compilations of best practices or guides to implementation. During Kranich's presidency, the ALA did produce brochures with suggestions and examples of civic library services. However, a recent sampling of the organizations cited by ALA demonstrates how quickly that information is outdated, and reveals that with a few exceptions these efforts were short-lived.

As for formal training at the master's or doctoral levels, there is little content to suggest that civic or even community librarianship has a pedagogical base. Aside from isolated instances reflecting the interests of an individual professor, no curriculum framework promotes new librarians' capacities to carry out their civic roles. Students invariably learn about the democratic "values" on which the public library is based and key concepts such as freedom of expression and the "right to know," but have no course in which they might discuss the implications for those values in terms of functionality, professional services, and allocation of resources. Graduate programs offer courses on community information networks, digital communities, community information systems, and outreach to local communities, but almost nothing that offers the background required for new professionals to actualize the civic value of their library.

The picture is even bleaker in professional development and continuing education. Whereas workshops on electronic databases and digital reference seem to proliferate as fast as the growth of information, very few workshops or trainings on civic librarianship exist in the usual vehicles for professional education: professional meetings, workshops offered by state library development offices and library systems, or the like. In recent months, there have been only two civic-oriented opportunities. The first was a workshop offered at an ALA meeting in San Diego, for training librarians to become moderators of public-issues forums organized through the Kettering Foundation and the National Issues Forum. The other was a workshop, "Libraries and Communities: Fostering Civic Engagement," hosted by the University of Illinois in connection with Professor Ann Bishop's efforts to develop Community Inquiry Labs and Community Inquiry Specialists capable of doing community research. These are the exceptions rather than the rule.

IV. Need for a Service Model

If we ask why the profession has given mainly lip service to something that, in theory, ought to be one of its basic functions, perhaps one answer is that the concept of the civic library has lacked a clear outline and mechanism by which it may be discussed, made an operational reality, or evaluated. The most recent analysis of the library's civic aspect is Ronald McCabe's *Civic Librarianship*, published in 2001, which thoroughly traces the strands of thinking about libraries as social and civic institutions, and discusses what he sees as two competing professional perspectives, the "libertarian" and the "communitarian." He argues that the libertarian perspective, which emphasizes the individual patron and personal rights, has undermined the profession's traditional commitment to its fundamental educational and civic mission. Calling for a restoration of the library's "democratic social authority" through a focus on its capacity for building communities, he offers the following "working definition": "Civic librarianship seeks to strengthen communities through development strategies that renew the public library's mission of education for a democratic society." McCabe recommends "strategies for action," including "promoting community identity, dialogue, collaboration and evaluation," but does not try to imagine what a civic library would look like (pp. 32ff, 85ff, 77, 79, 81).

The lack of a practical framework for bringing the Civic Library into action may be rooted in the fact that leading library educators and theorists are interested in different civic aspects of the library and do not ordinarily look at these aspects as part of a coherent whole. For instance, there is a very significant body of research and advocacy for the role of libraries as local information organizers and aggregators within the larger movement to create community information networks. Joan Durrance, of the University of Michigan School of Information Science, and her colleague Karen Pettigrew have documented the evolution, organization, content and impacts of community information networks and analyzed those with strong roots in public libraries.[1] Their perspective is informed by a special focus on the role of libraries as community institutions within emerging digital information systems. Their work is unusual in that they have worked closely with some public libraries to foster new practices in the realm of library-community information networks. The Flint Community Networking Initiative, referenced below, has benefited from Durrance's efforts to partner with the Flint Public Library.

The Durrance and Pettigrew perspective is not antithetical to, and in fact operates parallel with, another perspective focusing on the broader definition of the library as a community center and a key institution in the larger movement to "build communities." Sarah Long and Ron McCabe, mentioned earlier, and Kathleen de la Peña McCook are among those whose writings and teaching relate libraries to the communitarian movement

---

[1] See, for example, Durrance's bio on the University Michigan website, www.intel.si.umich.edu/cfdocs/si/courses/people/faculty-detail.cfm?passID = 32; and the "Library Highlights for 2002" page of the IMLS website, www.imls.gov/closer/archive/hlt_l1202.htm.

and the general notion of community building. Their leadership has advanced professional trends with respect to the provision of public spaces and functions that promote social interaction and community identity and, at the same time, have promoted community-library partnerships and collaboration. McCook's online newsletter, *A Librarian at Every Table* (McCook), has helped foster awareness and understanding of community building as an important function of the community librarian.

Another parallel and important aspect of the civic library discussion has been led by librarians, including Nancy Kranich, concerned about access to information and the library's role in ensuring information equity in the digital age. As the library community has struggled to grapple with the profound impacts of new technologies, including, but not limited to, issues of intellectual property, privacy and the commercialization of information, they have also led efforts to protect the public's right to government and other information. This work has underscored the continuing importance of the library from a democratic perspective, and has led to renewed appreciation within the profession of the inherent importance of the library as a vehicle for actualizing democracy through ensuring access to information.

Some library theorists are working on new roles for libraries, roles that may extend traditional functions or that may lead to new opportunities. One of these is the movement to re-think the connecting role of the library, broadening its functions from connecting people to information to connecting people to community service and work opportunities. They see libraries as important community-based vehicles for citizen engagement. Another of these is the movement to use libraries as a platform for engaging communities in the design of new community information systems, both virtual and actual. Ann Bishop and her colleagues at the University of Illinois School of Information are taking the lead in this work, some of which is organized around the concept of Community Inquiry Labs, noted above, which links community activists with public libraries in new ways that engage stakeholders in redesigning the institutional and electronic links between members of a community.[2]

These various approaches to viewing the library may vary in their theoretical aspects but are all rooted in the same philosophical perspective, which sees libraries as essential to democratic culture. In this respect they are all part of a larger whole—the whole civic library. What the approaches lack is a framework for connecting and institutionalizing the pieces as a comprehensive whole called the Civic Library.

---

[2] For Ann Bishop's work see, as one example, the web page of the Paseo Boricua Community Library Project, www.inquiry.uiuc.edu/cil/out.php?cilid = 1.

V. The LFF Civic Library Model

Perhaps in our democracy, the citizens, including professionals, need to be reminded that they speak democratic prose, to help them think more self-consciously about their deeply held values and assumptions. A formal model of the Civic Library that shows how its elements contribute to a democratic society might help sharpen the current discussion and enable librarians and others to begin talking in specific terms and thinking about transforming their thinking into action. In that spirit, I offer for consideration an approach developed by Libraries for the Future as part of its mission to highlight and strengthen libraries as essential democratic institutions.

Libraries for the Future was founded in 1992 by a group of citizens, including librarians, who were concerned, among other things, about the public's lack of awareness and understanding regarding the library's key importance for civic infrastructure. Its programs and projects have been nourished by the philosophy that libraries are core elements of democratic culture. Libraries for the Future advanced the concept of the Civic Library in 1995, to characterize an institution that is *self-conscious* about its civic role and *active* in efforts to promote community discourse, community identity and citizen participation. In 1997, I articulated the framework for a Civic Library in relation to the essential aspects of democracy: information equity, trust, dialogue and knowledge: "Beyond the dispensing of information, ideas or entertainment, beyond providing a place to read and think, the public library is also a civic institution. Indeed, the American public library both incarnates and furthers the mission of civic engagement." (Schull, 1997, pp. 11–12).

In 2000, my colleagues and I saw the need to help develop the Civic Library discussion in a more practical vein. At that time new studies by Robert Putnam and others were documenting increased declines in civic participation and social capital. We recognized both a need and an opportunity for libraries—a societal need for institutions that could promote democratic processes and an opportunity for libraries to be more intentional about their civic roles. In November 2000, we convened a working group of library leaders and others working on civic society to examine the concept of the Civic Library from a practical perspective (Libraries for the Future, 2000).

That meeting was a watershed. Not only did participants affirm the library's power as a symbol of democratic values but also its power as a real-time instrument for fostering democratic participation. They articulated the need for librarians to be more self-conscious of their assets as public conveners and educators, and more strategic in developing library programs and services, creating and re-creating library spaces, and functioning as leaders in their communities in ways that purposefully promote public engagement. In addition, they identified six areas where the library could engage (and some were already engaging) in efforts that could fulfill Archibald MacLeish's call to action in 1940, urging libraries to become "active not passive agents of the democratic process." (MacLeish, 2003). The six areas of activity advanced at the 2000 meeting have, with very little refinement, become the basis for the model for the Civic Library:

1. Public Space
2. Community Information as a Medium for Engagement
3. Public Dialogue and Problem Solving
4. Citizenship Information and Education
5. Public Memory
6. Integrating the Newcomer

Each of these brief phrases requires elaboration; however, what is most important in all of them is the presence of intentionality on the part of librarians. As I noted earlier, the library profession has often made gestures toward the importance of the civic aspect but has rarely sought to make it a practical reality. Therefore, the model is an attempt to purposefully animate the civic aspect. Libraries can emulate the model only if they consciously seek to adopt its elements, completely or in part.

(1) *Public space* refers to the physical and spatial aspects of the library as they affect public use and behavior, and underscores the preeminent value of the library as a place for common experiences. Usage patterns in libraries reflect the extent to which people are seeking the common experience. A welcoming social and civic space is a prerequisite for many of the other functions of the Civic Library, and, through its very existence, helps to build the trust and connections between residents that are essential to a healthy democratic culture. The sense of place, the concept of a "commons," is affected not only by the architecture of the library but also the design and use of its external spaces and how they relate to surroundings. Most examples take the form of dedicated areas on the main floor or next to specific departments, while others express the principles of the commons through the design of meeting rooms, study areas or the public areas. Many combine state-of-the-art technology and different study spaces for group projects, individual study or larger meetings (Beagle, 1999). The new Salt Lake City (UT) Public Library that opened in February 2003 was designed intentionally to function as a "commons for the city." A six-storey, walkable wall embraces the library plaza, creating an "Urban Room," with shops and services at ground level and reading galleries above. It is but one example of a library designed to foster social and civic interaction as a public good (Bagley, 2003).

(2) *Community information as a medium for engagement* touches on the library's ability to play a leadership role through organization, and management of local information and through creation of local information networks. A paradoxical result of new information technologies is that it has become easier for us to interact with people and activities across the globe than right across the street. Some observers of democratic culture believe this growing imbalance between global and local information is contributing to an increasing sense of disconnection in American culture and declines in civic participation. Libraries can help mitigate the downside of new technology by emphasizing community connections that are both face to face and electronic, and by organizing and disseminating local information for local residents. The public library can also use information strategically to build social capital by linking the skills and interests of residents to opportunities for service that benefit the overall community. In doing so, the library brings into play its many connections with other constituents of the community such as schools, museums, senior centers, environmental centers, health and childcare agencies, businesses, cultural institutions and youth organizations. As libraries evolve their civic services, they are doing so with input from residents—stakeholders—whose knowledge of their communities can ensure the relevance and appropriateness of the libraries' collections, spaces, services, partnerships and other civic functions.

The Saginaw (MI) Public Library's provision of online information resources illustrates how libraries are using networked information creatively to strengthen connections across the community—connections between people, people and services, people and other sources of local information. Saginaw Community Connection is an online database of over 800 local agencies and organizations; Saginaw Images provides online access to the history of Saginaw in photographs and essays; and Saginaw Facts and History provides a wide array of recreational, historical and practical information. In addition, the library has led development of a collaborative online community events calendar, GoSaginaw (Saginaw).

(3) *Community dialogue and problem solving* are essential for addressing important local issues and concerns. While all libraries can and do provide space for local organizations to carry out meetings or performances, the Civic Library may also organize the meetings and help the community set and examine its agenda. The same information networking noted above, as key for civic participation, also puts the library into a unique place in the community and gives it the opportunity to lead in the solution of local problems. In collaboration with local organizations, schools, and public entities, the library can convene and moderate public forums or stimulate discussion and debate on public affairs, local issues, or other matters that require public deliberation. Libraries fulfill the "forum" function in various ways. Some

provide forums using the National Issues Forums model or the Study Circles approach. Others host meetings and provide research assistance for residents who are tackling local problems. Still others are developing technological solutions to assist communities in securing local data, and engaging communities in inquiry-based approaches to shared challenges. In one example a group of libraries in Arizona worked with the Arizona Community Foundation, Arizona Humanities Council, and Libraries for the Future to organize and host a series of "community conversations" after the attacks of September 11, 2001. These libraries recognized that not only are libraries natural places to begin the public dialogue necessary in such a crisis, but that they should also take the lead in mobilizing their resources for this purpose. The project involved town meetings, local dialogues, an online community toolkit giving information on speakers, resources and reference material, and audiovisual support (Arizona, 2002).

(4) *Citizenship information and education* lie at the heart of civic life. However, citizenship and participation in the culture of democracy are learned skills, which newcomers may lack when they arrive; even many native-born residents may understand and practice them poorly or not at all. Public libraries would seem to be ideal learning environments, given their historic mission of welcoming all people, and their array of resources and services, including public affairs programs, films, and tutoring and mentoring programs. While it is true that all public libraries have information on citizenship and collections that support understanding of the meaning and functions of citizenship, it is also true that most do not reach out to those who may not know how to find these resources or how to start the process of citizenship. The Queens Borough (NY) Public Library's New Americans program is one exception.[3] Another is Citizenship through the Library, a program developed by the Ross-Barnum Branch of the Denver Public Library to engage Vietnamese and Latino adults and teenagers through readings and classes focusing on preparing residents for the citizenship examination (Denver Public Library, 2002).

(5) *Public Memory* refers to the library's responsibility to preserve and make accessible the records, images and other cultural artifacts that are meaningful to its audiences, especially as they relate to the historical and cultural experiences of local residents. Through this function the public library plays an important role in fostering community identity and a sense of a shared cultural heritage. This can come about through the use of new media to widen access to the vast array of local historical and cultural materials in local libraries, by building these collections through outreach to

---

[3]For the Queens Borough Public Library's "New Americans Program", see www.queenslibrary.org/programs/nap/index.asp.

nontraditional library audiences and newcomer groups, or by interpreting these collections and using them to better understand the local environment. Twelve years ago staff of the Los Angeles Public Library's photo archives became concerned that the collections did not include adequate materials representing minority families in Los Angeles, particularly newcomer families. They initiated a project, "Shades of LA," to collect pictures from family albums, through which images were chosen and copied for the library along with the story about the image provided by the donating family. The images included daily life, social organizations, work, personal and holiday celebrations, and migration and immigration activities. The strong public response enabled the library to expand its collection and gain valuable new knowledge about the city's current residents. At the same time, the project enabled those individuals and families who contributed photos to believe that they, too, were represented in the story of the city and in the city's library collection.[4]

While libraries have traditionally been understood as repositories of community history and culture, in many situations this "memory" function has been more passive than active. The active collecting going on today, through oral history, photo documentation and intergenerational exchange, are reinforced by new capacities to digitize and disseminate images and information—all part of the phenomenon of libraries reinterpreting their preservation function in the context of today's communities and today's technologies.

(6) *Integrating the newcomer* has long been a function of public libraries, made even more important by the growing influx of newcomers in recent decades. As the nation experiences the greatest wave of immigration in a century, libraries are exceptionally important locales for newcomer information and education. They are, increasingly, becoming vehicles for engaging newcomers in the community, from Phoenix, AZ, where the public library is collaborating with local refugee organizations to help connect families with services and institutions, to St. Paul, MN, where the library has established a special center for support of small businesses operated by immigrants and refugees (McCook, 2001). The Providence Public Library has created the "Cambodian Family Journey," a partnership with the Cambodian Society of Rhode Island that involves reading and discussion programs, oral history, bilingual conversations and other intergenerational activities to enhance communication between parents and teenagers and promote exchange between generations, cultures and languages. The project has enabled older members of the Cambodian community to share their

[4]For the "Shades of LA" program, see www.lapl.org/elec_neigh/index.html.

history and culture with young people—both Cambodians and other teens, and offered opportunities for adults and teens to sharpen their English and Khmer. Teenagers in the program have learned computer skills and designed web pages that include web films about their own family's unique story. The library has been able to expand its collection of important cultural information on new residents while also building the basis for an ongoing relationship with Cambodian-Americans in the city and the state. Cambodian-Americans are now actually participating in the library's governance and contributing to its operations and cultural perspective. Examples such as these suggest that the local library has a crucial role in helping newcomers become part of the civic and social life of the community.[5]

VI. Civic Libraries in Practice

The components of the Civic Library outlined above are alive and vigorous all around us. This entire essay could be devoted to cataloging the variations on the "commons" as an architectural design approach, or documenting examples of community information networks and citizenship education being carried out in libraries from Pelican Rapids, MN to Miami-Dade, FL. However, the significance of these isolated services in terms of a particular library or the generic "library" cannot be realized without clarifying their connection to one another as pieces of an intentional whole, i.e., the Civic Library. Those that are conscious of their leadership as civic institutions, those that carry out more than one of these functions, and those that demonstrate a commitment to institutional engagement in relation to the issues of the nation and of their local communities are the libraries that offer new models of the Civic Library. In preparation for this essay, I did a quick national scan of public libraries that have, on their own initiative, developed components of the Civic Library and, as such, exemplify an emerging trend.

The scan found a handful of libraries that are now functioning as civic libraries in everything but name. The six case studies below illustrate how some in the library community see a need for libraries to do business differently, to move beyond collecting, organizing and providing books and information, beyond provision of meeting rooms and reading clubs, beyond face-to-face and virtual reference, to stimulate public discourse and encourage public participation. These librarians understand their roles as leaders in the emerging information chain, and as essential instruments for

---

[5]For the Phoenix Public Library, see www.lff.org/programs/ircgrant.html. For the St. Paul Public Library, see McCook (2001). For the Providence Public Library, see www.geocities.com/cambodianjourney and www.provlib.org/community/events/camlit.htm.

framing and examining public questions. They are not only "at the table"—they are shaping the table.

A. Virginia Beach Public Library

> We are fortunate to be operating in the context of the city-wide planning, whereby the contributions we can make to improving the city are visible and desired. What is happening here demonstrates the role that libraries can and should play within communities. It is founded on the premise that the library is a place of public deliberation. The assets we bring to this process are many: we are neutral; we already have great credibility with the public; we can offer backup research and analysis for any subject under discussion. We believe the civic role of the library is basic to what we are and we welcome the opportunity to develop this role in as many ways as possible.
> (Mary Sims, Director of the Virginia Beach Public Library)

Virginia Beach, a small city in Virginia, has invested heavily in a 10-year strategic planning process involving all city agencies and many residents. One of the key concepts informing the planning has been "A Community for a Lifetime," the vision of a city where the voices of citizens would be heard, government would be responsive, public agencies would work together to share resources and ideas, and citizens would have increased opportunities to meet together and to discuss issues of common interest. The public library was heavily involved in the process of developing this concept, with the director heading up one of the seven planning teams, Quality Education and Lifelong Learning. Library managers were involved in monthly meetings with the City Manager to learn about how to operationalize this concept across the city's services.[6]

The Virginia Beach city planning process has given the library a stronger civic role. The library has been designated as the city's primary catalyst and facilitator for public affairs discussions and debate, and maintains an electronic forum for residents on the city's web page. Its staff are part of citywide training in customer service and partnerships, with some being trained as moderators or facilitators of public forums.

Befitting the library's new place at the center of a civic effort, its six branches are scheduled to be outfitted in ways that promote discussion and dialogue, as they come due for renovation. The Princess Anne Branch is the first to be designed to reflect the new emphasis. The announcement of its opening last fall referred to it as "The Library of the Future." The spaces, furniture and staffing plan are all designed to raise the level of service to the community and convey a strong aesthetic sense about the library.

[6]The profile of Virginia Beach Public Library is based on an interview with the director and information from the library's website.

Immediately inside the front door is an area designed to encourage the sharing of community information and the exchange of viewpoints; chairs are arranged so as to promote social interaction and discussion. A spacious meeting room will become a chosen location for public dialogue about issues of importance to the surrounding neighborhoods. Information providers move about, looking for patrons with questions or a child needing an impromptu storytime. One staff member described the new philosophy of customer service outreach as "approachable. They can approach us and we will approach them and offer our help."

Beyond the redesign of the library spaces, the use of the library for research on public issues and the development of public librarians as forum facilitators, the Virginia Beach Library is continuing to refine its work as a civic institution. Specifically, it is considering a proposal advanced by the City Council to develop a Civic Academy, in one or more of its facilities, as a place for learning about public issues taking part in civic dialogues. The academy's mission would be: "To fashion a sustainable way for the community to plan and engage in informed public deliberation of issues of importance to citizens, to foster civility in public discussions and to promote understanding of public policy issues." According to Library Director Sims, the academy would naturally extend the library's efforts to integrate civic library services and programs.

B. Carnegie Library of Pittsburgh (PA)

> People are thirsty to talk about issues and the library is the natural place for them to get informed on the issues and debate their questions. This kind of activity is every bit as important as a summer reading program. We must develop approaches and staff to carry these activities out so that we can say. This is what libraries do.
>
> (Herb Elish, Director, Carnegie Library of Pittsburgh)

As a director of a major urban library, Herb Elish is highly conscious of local needs for information—and of the need for opportunities to explore the meaning of that information through public forums, polls, online discussions and community dialogues. He believes that libraries "can help respond to a very important societal need that is not being fulfilled anywhere...libraries are well positioned to take on the public forum role. Normally libraries just provide material, but today it is important to help people think and listen and learn and discuss the important national and local issues."[7]

Acting on this conviction, Elish chose to host one of a series of "Citizen Deliberations" that took place in communities across the nation in January

---

[7] The profile of Carnegie Library of Pittsburgh is based on an interview with the director and material from the library's website.

2004. The deliberations were developed by MacNeil/Lehrer Productions, as part of *By the People: America in the World*, and drew about 1000 participants nationwide. The Carnegie Library was the only metropolitan library in the nation to host a poll. It partnered with Carnegie Mellon University, the World Affairs Council of Pittsburgh, the National Council of Jewish Women of Pittsburgh, and other organizations.

Deliberative Polling is a process designed to reveal how much people change their opinion after they have gained new information and insight about an issue. In the case of the Carnegie Library event, which drew about 100 residents, who were considered a representative sample of the community, participants were asked their opinions on selected key issues, such as whether the Iraq invasion had interfered with the national effort against terrorism. Then they were given background information on the issues and asked to meet in small groups to debate and frame a position, after which they cross-examined a panel of experts and then discussed the issues again in small groups. Finally, they were asked their views again on the issues.

The results were remarkable, according to Elish, who attended the event and was struck by the extent to which people changed their views as they gained new information and insights throughout the day. "Something happened in the group discussions that made people more tolerant of one another at the end of the day. Perhaps that can happen especially easily in a library, which has a climate that makes it more possible for people to listen to one another and learn. We need to understand the process better and especially the value that is added by the library." Elish wants to integrate this new polling function into the ongoing activities of the library, and he is encouraging new approaches to space design, particularly for the renovation of the central library, to strengthen the library's capacity to promote social and civic interaction and to be more visible as a central community place.

The Carnegie Library offers a number of other services that distinguish it as a leader in civic practices. The Three Rivers Free-Net is one of the first and most extensively developed community information systems in the country, and also one of the first and still among a handful to have been developed and managed by a library. According to Joan Durrance and Karen Pettigrew, experts on community networks, the fact that the Three Rivers Free-Net was created by a library and embodies the perspectives of librarians makes it especially effective in relation to other local information networks. It is easier to use, for example, and more reliable. The Free-Net hosts the web pages of local nonprofit organizations, offers extensive training, and hosts listservs that are better organized and easier to follow than on most community networks.

These services promote the flow of information among southwest Pittsburgh organizations and individuals. As noted in Durrance and Pettigrew's *How Libraries and Librarians Help*, the Free-Net is not a "passive portal," but a natural extension of traditional library functions that has become a vehicle for community building by helping to unify residents around areas of interest and concern.[8]

In numerous other aspects the Carnegie Library reflects the potential of libraries as civic and cultural integrators, engaging people in a wide array of community services and opportunities. In association with Pittsburgh Arts and Lectures, the Carnegie Library offers regular series of readings, lectures and book-related events to promote the written word and exchange of ideas. In terms of the "memory" function of the Civic Library, the Carnegie Library has collected significant numbers of images that document various aspects of the social and labor history of Pittsburgh. Again, these are not stored passively; rather, many are organized thematically and presented as virtual exhibitions on the web.

C. Flint (MI) Public Library

> If the public library, our most neutral ground, cannot offer a venue to talk about what is on people's minds, then who can? The library seeks to shed light but not heat on public debate.
> (Gloria Coles, Director of the Flint Public Library)

The Flint Public Library operates as a Civic Library in the sense that so many of its activities are similar to those outlined as components of the Civic Library model. These include its efforts to preserve and interpret the social history and cultural traditions of the diverse communities it serves, provide a common ground for the examination of issues affecting the quality of life in Flint, and build the capacity of other organizations through improving skills in the use of technology and information applications.[9]

One important example is the effort to help local organizations leverage new technologies and networked information. Through the Flint Community Networking Initiative the library aims to improve the way communities access and use information, including the provision of space, equipment and assistance to help community-based organizations become skilled in the use of online resources and technology. Started in 1995 as a cooperative project with the University of Michigan School of Information Science, the Networking Initiative began with the Flint Public Library Internet Training and Community Networking Center Lab, which sponsors presentations and speaking engagements that show community leaders how to use new technologies in ways that are meaningful to Flint. The Networking Initiative has helped many community organizations develop their capacities online,

---

[8] *Three Rivers Free-Net: Free to the People* www.si.umich.edu/libhelp/TRFN_profile.htm.

[9] The profile of Flint Public Library is based on an interview with the director, material from the library's website, and brochures.

including the design of a website for the local school district. A valuable skill-building effort is WebStudio, a physical gathering of people in the Flint Public Library training center who learn together how to create community information for the web. The goal is to increase the quantity and quality of community information available online by giving those who have the information skills to share it.

Although the Flint Community Networking Initiative strives to make geographic place less important, the training lab has reinforced the idea of the library as a place. Increasingly, the Flint Public Library is being thought of as the place to go to learn about information technology, experience it, use it, and meet others who support those efforts.

The Flint Timeline Project, which presents two centuries of images and stories of Flint on the library's web page, is another use of new technologies to strengthen the city's understanding of itself. Created by Flint Public Library staff and members of the community, with support from the University of Michigan's School of Information, this dynamic resource is a model for collaborative, community-oriented projects. It presents a wide array of information on aspects of local history ranging from the automotive history of Flint to education, labor, music and industries. Programs and events organized by the library reinforce community identity and use of the Timeline's content.

The Flint Public Library has been a leader in public dialogue, exemplified by its four lecture series, presented under the rubric "Race and Racism" and supported by a grant from the Charles Stewart Mott Foundation, which presented a variety of views on the current state of race relations in America. The Library's brochure for the first series, "Perspectives on Race and Racism," states that "the Flint Public Library will use its unique position as neutral territory and meeting place for all segments of the community, to present each of the authors in a two hour forum. The presentations will be based on the author's recent books. Following the formal lecture three area residents will add a local perspective to the dialogue. The evening will conclude with discussion and a question and answer period in which everyone is invited to explore, clarify and redefine their own views and attitudes on race and racism." Subsequent lectures series were entitled "Race in the 21st Century, A Continuing Dialogue," "Issues of Race," and "Beyond Barriers." The library billed these programs as Community Conversations, and the brochures all carried the phrase "Creating Racial, Cultural and Religious Understanding."

During its 4-year run, "Race and Racism" had a strong community impact. The four series gave momentum to the development of an anti-racist organization called FACTER (Flint Area Citizens to End Racism), housed at the library since hiring its first project director in 2003. As another example, following the first lectures in 1998–1999, area residents asked the library to organize a book club in which they could tackle racism using works of fiction. Each month, the Fiction Only Club, now in its fifth year, draws 10–20 participants, who come from all parts of town and from different economic and social backgrounds.

The Flint programs offer an important model of civic librarianship, in which a library director recognizes the power of the library as a place of dialogue, problem solving and learning, and invests time and resources to provide a venue and a social and intellectual context for the discussion.

D. New Haven (CT) Free Public Library

> New Haven Free Public Library provides free and equal access to knowledge and information in an environment conducive to study and resource sharing. Through its collection, media, services and programs, the library promotes literacy, reading, personal development and cultural understanding for the individual and the community at large.
>
> (New Haven Free Public Library mission statement)

"The community at large" is key to understanding the New Haven approach. Through a wide array of partnerships with local schools, arts organizations, service groups, city agencies and nonprofit organizations, the library functions, and is perceived as, a hub for information, cultural exchange, lifelong learning and community development. The Director, James Welbourne, has been a leader in the public library movement to develop community information services, and in previous positions in Baltimore and Pittsburgh emphasized collaboration as means of applying library resources to promote community development and foster equity of opportunity. Under his leadership New Haven is moving in the same direction. Four initiatives currently underway exemplify the library's efforts to multiply local assets through collaboration.[10]

First, in the area of services for children, youth and families, the New Haven Free Public Library is a leader in the integration and coordination of programs that foster emergent literacy and positive youth development. It offers Family Place Libraries, a program that emphasizes outreach to families and caregivers with very young children and links families to information, to one another, to developmentally appropriate learning resources, and to other community services that enhance healthy development. The library's Readmobile, developed cooperatively with the local school district to serve failing schools and underserved neighborhoods, also reflects the commitment

---

[10] The profile of New Haven Free Public Library is based on an interview with the director, material from the library's website, and a report by New Haven Free Public Library about its participation in the "Coming of Age Forum," December 2003.

to pairing resources with others for maximum community benefit—a fundamental characteristic of the Civic Library.

New Haven's Non-Profit Resource Center (NPRC), initiated in 1998, is a Cooperating Collection of the Foundation Center and an Affiliate Library of the Association of Fundraising Professionals through which the library provides a comprehensive collection of materials to assist nonprofit organizations in fund development and nonprofit management. The NPRC serves grant seekers, grant makers, researchers, policymakers, board members, volunteers, consultants, fundraisers and others associated with nonprofits in the greater New Haven area. The library not only provides information to strengthen the nonprofit sector but also conducts workshops, convenes groups, and even works with smaller libraries in the region to ensure that they know how to make the collections accessible to their local nonprofit organizations. Far from being a passive resource, the information is the starting point for intentional outreach to targeted constituencies.

The library is also improving life in New Haven through a strategic partnership with the Volunteer Center of Greater New Haven. This partnership may enable the library to develop a more systematic, community-wide approach to informing, connecting and mobilizing volunteers. The library and the Volunteer Center recently organized a Community Conversation with adults approaching or just starting retirement, through which both institutions learned more about the needs and interests of adults who want to serve their communities in retirement. The library is exploring how it might be a better place for the civic engagement of older adults, and has already begun testing this theory with the Ben Carson Reading Club, a reading promotion effort that engaged Yale alumni in outreach to the schools, churches and other community organizations. Welbourne says that the experiment enabled the library "to envision a variety of ways to utilize volunteers and to expand its commitment to reading promotion throughout the school year."

The New Haven Free Public Library's work with local cultural organizations is yet another reflection of its commitment to community-building through collaboration. According to Welbourne, "a vital city includes a dynamic cultural community, and the library can be instrumental in both building cultural capacity and facilitating public access to the arts." One manifestation of this role was a recent performance at the library organized by the Open End Theatre, a group that reaches out to youth through dramatic performances in which young people in the audience are asked to comment and make choices. One event, held in the library's predominately African-American Stetson Branch, focused on the sensitive issues of AIDS and Teen Pregnancy. A second event, a dramatic multi-media performance on the Tuskegee Airmen, drew an audience of 250, and turned the library into a venue for theater and dialogue, expressing vividly a commitment to work with the community on positive youth development through the arts.

The fact that the librarians themselves do not put "civic" labels on the activities sketched above does not diminish the civic quality of what is happening at New Haven Free Public Library; rather, it is an artifact of professional training, which lacks a coherent model for understanding what the librarians are doing on a daily basis. While these activities may not, on the surface, be explicitly labeled as civic programs, it is easy to see how they express the core values of the Civic Library, such as the importance of public space, community identity, community networking, or community problem solving.

E. Cleveland Public Library

> The mission of the Cleveland Public Library is to be the best urban library system in the country by providing access to the worldwide information that people and organizations need in a timely, convenient, and equitable manner.
> (Cleveland Public Library mission statement)

The largest of the profiled libraries, Cleveland Public is known nationally for its achievements in applying new technologies to make networked information and services available to the widest possible number of people and organizations. The library is also a leader in developing services at the neighborhood level that ensure equity of access and promote community participation. For example, through its mobile "People's University" the library is reinventing the traditional bookmobile to promote access to information and ideas. Many activities and services could be identified as fitting the Civic Library model. For example, the library carries out projects and services that promote community identity, help local groups solve problems, provide information to inform voters and integrate newcomers into the civic and cultural fabric of the city. For the purposes of this discussion I will focus on three activities that demonstrate the library's commitment to moving beyond business as usual and to reorganizing its traditional services to build community relationships and local participation.[11]

One initiative, *Yread*, illustrates libraries' growing efforts to use emerging information systems and web-based software to build connections between people. *Yread* is a website for teens, educators, and individuals who work with young adults and who care about access to literature and ideas. Funded by the Ohio Library Foundation's Drew Carey Young Adult Service Program, *Yread*

---

[11] The profile of Cleveland Public Library is based on material from the library's website and a proposal to Libraries for the Future for the MetLife Reading America intergenerational program, December 2002.

offers periodic live, online book discussions that reinforce the library's ongoing program of local appearances by nationally known authors, including well-known Hispanic authors Piri Thomas and Julia Alvarez, and Chinese-American author Maxine Hong Kingston. The *Yread* website has helped to integrate the perspectives of these authors and provide a forum for diverse points of view that cross the cultures represented in the city's population. It moves beyond the provision of information to promotion of interaction between generations, authors and readers and between readers themselves.

North Coast Seniors Connection is another example of the library's efforts to strengthen community connections through new uses of communications and information systems. Creation of this website (http://www.seniorconnect.org/(2ekfen551qin1l450f1f4f55)/Default.aspx), designed for older adults and those that care for and about them in Cuyahoga County, was identified as a priority by Cleveland's Senior Success Vision Council. The library developed the site as the first step in a grant-funded project to give seniors better access to information and encourage them to use computers to learn about opportunities in greater Cleveland. In addition to creating a website, the staff of North Coast Seniors Connection want to make presentations and small group lessons enabling older adults to "become information literate and interact electronically with the community." North Coast Seniors Connection is one outgrowth of the library's strategic plan, which was developed with community input. It demonstrates the library's active use of information and communications to strengthen links between people and in so doing to improve the quality of neighborhood life.

Another community-building initiative is "Many Voices, Many Lives" (MVML), a program that highlights the experiences of people living with AIDS and promotes cross-cultural discussion and participation on the theme of AIDS. A year-long project, it involves collaboration with other community organizations, including the AIDS Taskforce of Greater Cleveland and their SAMM (Stopping AIDS is My Mission) program, and is supported by the Metlife Foundation through Libraries for the Future's "Metlife Reading America program." The library explains that the project is creating "a year long opportunity to build relationships with target communities. The subject matter will lend itself to a level of intimacy that is not often achieved in relationship with library patrons."

In partnership with Cleveland's Playhouse Square Center, the library works with the collaborating agencies to select the main book for the community discussions, create lists of films and books, and prepare promotional materials. Through small group discussions held throughout the city, related activities such as film showings and performance workshops, and a final performance and community event celebrating World AIDS Day, the program unites the city in a discussion of a key public issue affecting the lives of many young people and their families. MVML involves outreach to Cleveland's teens and their parents, with a special emphasis on immigrant families and on reaching inner ring suburban libraries with clusters of new immigrant populations.

The theme of AIDS awareness and advocacy through literature and performance distinguishes the Cleveland Library's reading program from other "one book" efforts taking place throughout the country. Another distinguishing feature is the emphasis on youth engagement and youth leadership development. Teen advocates from SAMM assist in recruitment and take part as workshop facilitators and group leaders. The Library's commitment to taking the lead in helping to address a major public challenge is reflected in the program's stated goal, "to foster a high level of involvement and interdependence among participations and to raise the level of commitment."

F. Johnson County (KS) Public Library

> This is a new role for the library and for the public's perception of the library. As librarians we are trying to find ways to be more useful to the community, to package our skills and information and spaces to help promote dialogue and problem solving. Previously, people didn't think of the library as a place to go for examining problems.
> (Donna Lauffer, Associate Director for Branches/Facilities, Johnson County Public Library)

As at the Virginia Beach Public Library, the Johnson County Public Library's evolution as a Civic Library was stimulated by a government planning initiative. In 2002, the Johnson County Library Board of Directors adopted a strategic plan, "Connections—Enriching Lives, Building Community." One of the plan's goals called for the library to provide a venue for discussion of public issues. Simultaneously, library staff were starting a planning process, asking: How should we be looking at the library in the 21st century? What are we good at now? What could we be good at? What do we understand is useful for our community? To try to answer these questions the library undertook a public opinion survey and an environmental scan. The opinion survey revealed strong interest in the library's abilities to connect people and to help them make decisions. The environmental scan focused staff attention on the work of Robert Putnam and his studies arguing for a national loss of social capital and the need for building stronger community institutions and relations. The library concluded that it could and should play a stronger role in community building through the provision of forums and opportunities for discussion.[12]

---

[12] The profile of Johnson County Public Library is based on an interview with the associate director for branch services and material from the library's website.

The library examined models for public forums and community discussions, including Study Circles, League of Women Voter Forums, and the National Issues Forums operated with support from the Kettering Foundation. Donna Lauffer, an associate director of the library, attended presentations and training on the National Issues Forums. At the same time, Johnson County's Urban Center, Kansas City, had finished the *Citistates Report*, a study that led to a series of reports on issues important to the future of the city. The library, with support from the Kauffman Foundation, began hosting forums with discussion guides based on the five issue areas examined in the study. The objectives of the Community Issues 101 Forums are to provide a place for the free exchange of ideas in order to make informed and nonpartisan decisions, encourage citizens to understand and envision fresh solutions for a stronger community, and give participants a clear understanding of the issues and the means to draw their own conclusions. Forums take the form of deliberative dialogues, in which groups move toward a public decision by debating different perspectives on issues and develop a consensus about solutions. The library's commitment to the forums is reflected by their release of staff to attend trainings, become forum moderators and facilitators, and work on the preparation of briefing materials and on marketing the programs.

To date, the Library has held five forums, on topics that are "timely, significant, and contributive to the long-term benefit of the community," and which "are supported by the library's bibliographic resources, support intellectual freedom principles," as well as "complex issues that merit exploration or further explanation." Topics examined to date are: Economic Development, Transportation, America's Role in the World; Educating Kansas Children; Race Relations: Beginning the Conversation.

According to Lauffer, "the Community Issues 101 Forums give Johnson County residents a chance to discuss and debate important issues that affect us all. These forums have brought people together that would never be together otherwise—there is a special value in having these discussions in the library that we are trying to better understand." Research conducted by the Kettering Foundation suggests that the programs will be "continually positioning the library as the nexus, offering space for ongoing dialogue and problem solving and linking people to people and people to information in support of community engagement." Lauffer is considering ways to refine the National Issues Forums model to ensure its relevance to local audiences and, in particular, those who may not be traditional users of the library.

As the experience of the Johnson Library points out, libraries and librarians can play a crucial role in reinvigorating civic engagement by convening and moderating public forums, providing research assistance for tackling local problems, and engaging communities in inquiry-based approaches to shared challenges.

VII. Conclusion

What will it take to realize the civic value of the library in a practical sense? What will it take to develop a new paradigm for service within the profession that can foster social capital and strengthen civil society? What will it take to foster capacities for transformation?

In advancing the concept of the Civic Library, we must first acknowledge the variations on civic practice that exist today, and find a way to unify them in a holistic model, such as the one proposed above, that can provide the basis for training, advocacy, policy development, funding strategies and evaluation. Furthermore, we will need to create systematic methods to expand practice beyond the work of a few pioneers and make it part of the library mainstream. Most of the library directors who are trying to make their libraries a more integral part of the civic fabric believe that they are acting alone, in response to unique local conditions, and that what they are doing may be uncommon in the library profession but must be done, nevertheless, because it is so important to the community. We need to validate what they are doing, by incorporating their stories into the professional narrative and making them widely known. We also need to acknowledge the pioneers as leaders of one of the most important imaginable practices that could exist among public librarians, and make them role models for others. Additionally, we will need to weave the concepts of the civic library into the academic training that prepares librarians for their career, and the in-service training that enables them to bring the civic library into their working vocabulary. At each library, the staff and the trustees will need to be sensitive to the library's civic potential, and conversant with the basic concepts and practice of the civic library.

Beyond purely professional aspects, we will need to educate the public, especially key decision makers and opinion makers, about the library's enhanced standing as a civic institution, ready to act as a partner in advancing civic dialogue and problem solving. Both directly and indirectly, we must foster new understanding regarding the library's strategic importance for the development of social capital and healthy, interactive communities. I will, therefore, end my analysis with a challenge to the library community, namely, to join in a planning effort to align rhetoric, potential and practice for the development of the Civic Library as a model for 21st century participation.

References

Arizona September 11th Community Conversations (2002). A report by Libraries for the Future and Arizona Community Foundation.

Bagley, W. (2003). Salt Lake City's library movement has a colorful and controversial past. *Salt Lake City Tribune*. February 2.

Beagle, D. (1999). Conceptualizing an information commons. *Journal of Academic Librarianship*. March, accessed via Wilsonweb.
Block, M. (2003). How to become a great public space: want to attract more users? Fred Kent and Phil Myrick of the project for public spaces suggest you look to your welcome mat. *American Libraries*, 72–76. April.
Demas, D., and Scherer, J. A. (2002). Esprit de place: maintaining and designing library buildings to provide transcendent spaces. *American Libraries*, 65–68. April.
Denver Public Library (2002). Citizenships Through Libraries Program. Report to Libraries for the Future, supported by Libraries for the Future through the MetLife Foundation Reading American Program, December.
Elish, H. (2003). Quoted in a press release, December 17, 2003, on the Carnegie Mellon University media relations website: www.cmu.edu/PR/releases03/031217_townhall.html.
Kranich, N. (ed.) (2001). *Libraries & Democracy: The Cornerstones of Liberty*. American Library Association, Chicago, IL.
Libraries for the Future (2000). The Civic Library: A Discussion on the Civic Dimensions of the Public Library and Possibilities for Strengthening Its Civic Role, an unpublished report of a meeting convened by Libraries for the Future in New York City on November 20, 2000.
MacLeish, A. (2003). Speech is reproduced on the website of Greenfield Community College, www.gcc.mass.edu/library/resources/MacLeishSpeechMay03.doc.
Malouf, M. B. (2003). SLC library challenges Utah architecture. *Salt Lake City Tribune*. February 9.
McCabe, R. B. (2001). *Civic Librarianship: Renewing the Social Mission of the Public Library*. Scarecrow Press, Lanham, MD.
McCook, K. A. (2001). Collaboration generates synergy: Saint Paul Public Library, the College of St. Catherine, and the 'Family Place' Program. *Reference & User Services Quarterly* **41**(1), 19–23. Fall.
McCook, K. A. (ed.) *A Librarian at Every Table*, www.cas.usf.edu/lis/a-librarian-at-everytable.
Molz, R. K., and Dain, P. (1999). *Civic Space/Cyberspace: The American Public Library in the Information Age*. MIT Press, Cambridge, MA, pp. 14–15.
Morris, J. (2002). The college library in the New Age. *University Business*: **5**, 27 October.
Nelson, S. S. (1998). *The New Planning for Results: A Streamlined Approach*: Public Library Association, Chicago, IL.
Putnam, R. D., and Feldstein, L. M. (2003). *Better Together: Restoring the American Community*. Simon & Schuster, New York.
Saginaw web page, www.saginaw.lib.mi.us/Community_Info/community_info.html.
Schull, D. D. (1997). A case sfor empowering the civic library. *Virginia Town and City*. **32**(3), 11–12. March.
Trustees of the Public Library of the City of Boston (1852). J.H. Eastburn, Boston, Available at www.scls.lib.wi.us/mcm/history/report_of_trustees.html.

Libraries and Learning

Robert S. Martin
Institute of Museum and Library Services, Washington, DC, USA

Libraries are social agencies. They exist to serve certain specific needs in our society. Changes in the environment in which libraries operate—in the technological infrastructure through which we deliver services, in the economic substrate that finances operations, in the social landscape that defines the communities that libraries serve—dictate corresponding changes in the way libraries structure and deliver services. One additional change in our environment—our emerging understanding of the nature of learning and the way learning interacts with other aspects of our environment—is likely to result in an even more rapid change in the coming decade.

Libraries—and librarians—have been coping with constant and rapidly accelerating changes in these environmental factors for the past three decades. Those changes have already had a dramatic effect on the resources and services that libraries provide. Whether one views these changes as evolutionary or revolutionary is largely a matter of semantics. Stephen Jay Gould posited a view of evolution in the natural world that he labeled "punctuated equilibrium," in which long periods of relative stability are interspersed with brief periods of very rapid change. It appears that libraries, in response to rapid changes in our environment, are undergoing such a period now. Our equilibrium has been punctuated.

The Institute of Museum and Library Services is an independent federal agency that serves as the primary source of federal grants for the nation's libraries and museums. Our grants to museums and libraries build institutional capacity, support core library and museum services, encourage excellence, foster collaboration between and among museums and libraries, and promote innovation. Through its grant programs and its convening authority, IMLS provides leadership for the library and museum fields. The National Leadership Grants program, in particular, has funded digital projects in both libraries and museums, which have dramatically enhanced public access to rich cultural heritage resources. National Leadership Grants have also fostered a culture of collaboration between and among museums and libraries, demonstrating the common mission of these important social agencies.

At IMLS we are convinced that museums and libraries are fundamentally social agencies that share the role of providing the resources and services that stimulate and support learning throughout the lifetime. In other words, we understand that museums and libraries are both agencies of public education, fundamental to the education infrastructure of our society. That simple recognition underlies the action of Congress that created IMLS in its present form less than a decade ago. That is why at IMLS we are dedicated to the purpose of creating and sustaining a nation of learners.

We often hear it said that today we are living in an information age. But in a world drowning in information, we are hungry for knowledge. That is why today, in the 21st century, we must be more than an information society. We must become a *learning* society.

A learning society requires that we do more than develop the hardware, software, telecommunications networks, and other services and systems that supply and organize content. It requires additional structure and context to enable learners around the globe to put knowledge to good use.

As Falk and Dierking (2002) have pointed out,

> Lifelong learning, long a utopian educational goal of our society, is increasingly becoming not just a necessity, but also a way of life. As our society is increasingly inundated with information, each individual finds it necessary to develop better strategies with which to analyze the increasing quantity of information in order to select that of high quality and broad utility. In a myriad of subtle and not-so-subtle ways, this necessity has resulted in America becoming a nation of lifelong learners.

Learning today is changing. What we know about learning—where, when and how it happens—is changing. Over the past 20 years, there has been an explosion of neuroscientific research. We now know more than ever about how the human brain learns, from infancy throughout the senior years. We now know that, although there are intense periods of rapid brain development in early childhood, our brains continue to develop and form new connections throughout our entire lives. We know that children are never too young to learn, and we know that lifelong learning extends the quantity and quality of life. Studies demonstrate that the capacity to learn increases at age 50 and extends well into the senior years.

The structures we have in place today for providing public education evolved in response to specific environmental conditions and social needs. They are largely an artifact of the late 18th and early 19th centuries. How else do we explain, for example, the persistence of the 9-month school calendar, which was developed in response to the needs of a rural agrarian society, in which three quarters of the population were engaged in farming? Today we live in an urban post-industrial society, in which only 4% of the population is engaged in agriculture, and yet the facilities we have created at great expense to house our schools and colleges sit fallow and unused one quarter each year.

Our current structures for learning—the formal education system embodying both the K-12 school system and the post-secondary education system—are relatively late developments in the evolution of human society.

It has only been within the past 100 years or so that the words *learning*, *education* and *schooling* came to be treated as synonyms. In fact, America has always drawn from many different sources, including but not limited to schooling (Falk and Dierking, 2002).

In short, it is only within the last century that we have come to believe that the best way—indeed, virtually the only way—for people to learn is in structured, formalized education systems. As Daniel Pink says in his challenging essay "School's Out,"

> Through most of history, people learned from tutors or their close relatives.... Not until the early 20th century did public schools as we know them—places where students segregated by age learn from government-certified professionals—become widespread. And not until the 1920s did attending one become compulsory.... Compulsory mass schooling is an aberration in both history and modern society. (Pink, 2001)

Today we are witnessing conspicuous challenges to the basic assumptions of schooling. The dramatic rise of home schooling in the past decade is one example. (The term "home schooling" is in fact a misnomer, since the last thing that the practitioners of this form of learning is interested in doing is recreating a school in the home.)

In fact, in our society we learn in three different sectors. We learn in the school. We learn in the workplace. And we learn in the home and community. The last of these three sectors is now frequently referred to as the free-choice learning sector, underscoring that learners in this environment are motivated by individual needs and interests.

In the 21st century, environmental conditions mandate that the ability to learn continuously throughout the lifetime is essential. Accelerating change has become a way of life. To navigate such change, minimize risk, and participate effectively in civic affairs, all Americans need access to learning throughout their lifetimes. The importance of continuous learning, free-choice learning, lifelong learning, for economic vitality and for personal fulfillment alike, is beyond question.

This contention is reinforced by the recently published *2003 OCLC Environmental Scan: Pattern Recognition*, which notes that we now operate in a Knowledge Economy:

> ...in which technology and the knowledge on which it is based are central motors of economic growth. This means that a growing number of workers manipulate symbols rather than machines. And it means that human or intellectual capital—the knowledge that comes from education, training, on-the-job experience and workplace-based e-learning—is central to sustaining personal and organizational advantage. (De Rosa *et al.*, 2004)

This has important implications, among which is that "the ability to learn and to adapt to change is a central life skill. Learning is valued as a crucial coping skill in an environment of change and flexibility." (De Rosa *et al.*, 2004).

The OCLC Environmental Scan also brings our attention to three major social trends: Self-service, moving to self-sufficiency; Satisfaction; and Seamlessness.

Self-service is becoming the norm in virtually every area of human activity. Whether it is at the gasoline pump, in retail checkout lines at grocery stores and home-improvement centers, or buying commodities and services online, self-service has quickly become the norm. Some experts estimate that within 2 years 95% of American supermarkets will have self-service checkout options (Grimes, 2004). And of course, there has been a notable trend towards self-service circulation kiosks in libraries.

The economic benefits for retailers and service providers resulting from self-service operations are obvious. However, studies indicate that many individuals—especially younger ones—actually prefer self-service operations over dealing with human beings. Computer terminals are perceived as more reliable and faster (Licata, 2004). "Online banking and online travel activities have disintermediated the humans who used to be the gatekeepers and guides to these services," notes the OCLC study, "but self-sufficiency and convenience are prime drivers for the consumer." (De Rosa *et al.*, 2004).

People who use the Internet to acquire information, products and services profess themselves extremely satisfied with the results that they achieve. They find what they want or need when they want or need it. They may be unaware that higher-quality or more relevant results might be achieved by having recourse to an expert intermediary, and they do not apparently care.

While self-sufficiency and satisfaction are important to learning, and to structuring library services that support learning, the importance of seamlessness is crucial, and possibly the dominant trend for the future of libraries. According to the OCLC report, in today's society

Libraries and Learning 87

The traditional separation of academic, leisure and work time is fusing into a seamless world aided and supported by nomadic computing and information appliances that support multiple activities. (De Rosa *et al.*, 2004)

The report goes on to stress that this is particularly significant among young adults, noting that

The freshman class of 2003 grew up with computers, multimedia, the Internet and a wired world. ... Their world is a seamless "infosphere" where the boundaries between work, play and study are gone. Computers are not technology, and multitasking is a way of life. ...the lines between workplace and home are blurred. (De Rosa *et al.*, 2004)

While the report uses this analysis as a foundation to describe the kinds of seamless services that libraries need to develop, I think it is important to recognize the implications that go beyond our institutional boxes, and to contemplate developing a seamless infrastructure for learning across all the social agencies and organizations that create, maintain, and provide access to resources that support learning. In short, we need to adopt a bold new vision of learning.

The responsibility for learning is not and should not be the exclusive preserve of formal educational institutions. It is a community-wide responsibility. Lifelong learning should be a continuum—with formal and non-formal learning opportunities complementing one another. Learning does not start at the schoolroom door; neither does it stop at that portal either. It is and should be ubiquitous.

Addressing the needs of the free-choice learning sector is now more critical than ever. As more people become self-guided learners throughout their lifetimes, institutions such as libraries, museums, and public broadcasters, among others, can help to stimulate and meet their demands.

At IMLS, one of our major efforts is to foster collaboration. We believe that effective collaboration is *the* strategy of the 21st century. It is aligned with how we are thinking about our communities as "holistic" environments, as social ecosystems in which we are part of an integrated whole. The kind of collaboration we try to foster is simply a mature and reflective recognition of intersecting nodes of interest, activity and mission.

Naturally at IMLS, we are interested in fostering collaboration between and among museums and libraries. It is inherent in our structure, and mandated by our governing statute. But we also think it is imperative to reach out beyond the museum and library and to find nodes of intersecting interest and mission among other players in the community.

One of the potential partners in which we have the most interest at present is public broadcasting. There is a growing awareness that the changes that broadcasters are going through, due in large part to the impact of digital technology, lead to a recognition of a pending convergence with museums and libraries. That convergence in turn reflects a similar convergence of museums, libraries and archives, driven by a recognition that in the digital environment, the boundaries between these kinds of cultural heritage agencies are blurring.

Recently, at IMLS, we began hearing consistent reports from our grantees, indicating that in the digital environment, libraries are beginning to "behave" more as museums and museums are behaving more as archives. In the traditional non-digital environment, libraries organize their collections and present them for use in response to a user's specific need or inquiry. A user comes into the library and asks "what do you have on topic X." For example, "Show me everything you have on impressionist painting, on Native American ritual objects, on Paleolithic protozoa."

Conversely, museums traditionally organize selections from their collections in topical or thematic interpretive and didactic exercises we call exhibitions. A user comes into the museum and looks at what the museum staff has selected, presented and interpreted. A museum-goer would not normally come into the museum and say "show me all of your impressionist paintings, show me all your Native American ritual objects, show me all your Paleolithic protozoa."

In the digital environment, these behaviors are almost precisely reversed. Museums for the first time can present their entire collection, cataloged and surrounded with metadata, retrievable in response to a user's specific interest or inquiry. Examples of such access are common. The National Gallery of Art, for example, has mounted a web interface to its entire collection, searchable by artist, title, subject, provenance, or accession number, as well by an expanded search on any combination of artist name, key words in title, school, style, date of creation, medium, and/or short list of popular subjects (see http://www.nga.gov/collection/srchart.htm). The Freer Gallery of Art of the Smithsonian Institution likewise has made its complete collection available online, retrievable through a web-based interface (see http://web4.si.edu/asia/collections/search.cfm).

Libraries, on the other hand, now routinely organize selected items from their collections in thematic presentations that tell a particular story, and even call these presentations "exhibitions." There are many examples of this behavior in libraries—indeed, it is now so commonplace as to be considered routine. The Louisiana State University Library, for example, mounts online versions of the excellent exhibitions developed in its Special Collections division (see http://www.lib.lsu.edu/special/exhibits/elecex.html). The Texas State Library and Archives presents another approach, offering a range of thematic exhibitions from its collections (see http://www.tsl.state.tx.us/treasures/index.html).

It is important to note that the users of these digital collections do not care, and may well not even be aware, that the originals of the digital surrogates that they use are in a museum, a library, an archive, or some other kind of institution. They really do not care how you define your institution—they just want access to the "stuff."

In much the same way, we now have evidence that now, in the digital arena, broadcasters too have been undergoing a transformation that results in behaviors that are more like museums and libraries.[1] Formerly, we have been accustomed to thinking about broadcasters as providing access to rich educational resources, but in a strictly synchronous manner. If we wanted to enjoy the educational content that they provide, we were expected to tune in on Thursday evening at 8:00 to see the latest program on the rings of Saturn, on the explorations of Lewis and Clark, or on the plays of Shakespeare.

But increasingly now this broadcast content is no longer "broadcast" in the conventional sense. It is accessed through cable or satellite connection. And increasingly we can also access content online, downloading an entire program from a website.

What's more, new digital video recording devices like TIVO are transforming the way that audiences interact with television programming, enabling the "viewer" to capture the broadcast, retain it for use at a later time, retrieved and used at the convenience of the receiver.

Traditional synchronous access to broadcast programming is declining and asynchronous use is becoming the norm. "Broadcasting" no longer adequately describes what broadcasters do—it instead describes the technology that they formerly used to do what they do. And they have come to realize that "broadcasting" is not the essence of their business—it is creating and providing access to educational content and opportunities.

There is one other important transformation for broadcasters. In the traditional context, the programming that is made available at 8:00 PM on Thursday evening is typically 50 minutes of content. This represents really only an executive summary of hours of material that have been captured or created, and edited down to fit the available programming slot. But it is now common to make at least some of that additional material available to the user, via the broadcaster's website. We have all heard the instruction at the end of a show or segment that we can find additional information at a specified URL.

So broadcasters are now trying to find ways to organize and present for use the vast quantities of raw material, surround it with metadata, and make it retrievable in response to a specific user inquiry. In short, in the digital

---

[1]The author is indebted to David Liroff of WGBH Boston for this overview of changes experienced by broadcasters in the digital environment.

environment, broadcasters are behaving more like museums and libraries. There appears to be a pending convergence of libraries, museums, archives and broadcasters in the digital environment.

This convergence is not restricted to the digital environment. Networked digital information technology has simply lifted the veil that has obscured the basic fact that the silos into which libraries, museums, archives, broadcasters, and other developers and purveyors of learning resources and opportunities are ghettos of our own making. There is no natural law that distinguishes library from museum, museum from archives. On the contrary, the natural state of affairs—underscored by our common history—is that the similarities among such agencies are far more striking than the differences. The distinctions have only arisen in the recent evolution of human history, in response to specific technological practices that separated object from text and manuscript from printed codex.

If we can posit that librarians, archivists and museum professionals are not separate and distinct professions, but rather different facets of a single unified profession, we will find that our ability to serve the needs of our communities is strengthened. If we re-envision ourselves as public servants, charged with the responsibility for collecting and organizing the materials that document our rich and diverse cultural heritage and enhancing access to those materials for our citizens, we will find that we can reshape our practices, learn from each other, and better attend to our users.

As we move forward in this 21st century, we certainly do need to change our thinking about how to develop and structure library services. We need to evolve into agencies that focus not on collections, but the needs of the users. However, there is nothing really new about this. In Zweizig's (1973) Syracuse University dissertation, he noted that "Librarians have too long focused on the user in the life of the library. We need instead to focus on the library in the life of the user."

We do indeed need to develop facilities that recognize, embrace, and encourage the collaborative and social nature of learning. We must create learning environments that empower student learning, enabling them to turn information into knowledge. We must extend these lessons from the realm of the university to all levels of formal education, from the kindergarten to the research university.

Beyond this, however, we must also embrace the same principles for libraries of every kind, including the public library. We must recognize and embrace the social nature of continuous learning, free-choice learning, that lasts the length of the lifetime. Public libraries must be conceived as a learning environment, providing spaces that foster and support the individual learner, as well as learners in every imaginable form of social grouping.

But this is not enough. We also need to think much more broadly, to envision a seamless learning infrastructure, one that stimulates and supports learning for learners of all ages, in all circumstances. If we are truly to empower individuals to fulfillment, to enable them to maximize their human potential, become contributing members of the knowledge economy, and to participate effectively in civic affairs, then we must build a fabric of social agencies that facilitates continuous lifelong learning. Such a fabric must integrate all of our current social agencies that create, manage, and provide access to learning resources, including schools, libraries, museums, archives, broadcasters, and a host of other organizations and agencies.

Here are a few modest specific recommendations:

- To enhance access to learning resources of all kinds, we need to develop consistent and reliable mechanisms for creating digital resources. We already spend enormous sums for education, and for learning resources. In the US, in 2001, we spent roughly $500 billion, or $1780 per capita, on education. That is 4.8% of GDP. In the same year, we also spent about $12 billion on libraries, or about $43 per capita. We think that about $50 billion is a reasonable approximation of what we spend on museums. Some of these resources need to be directed toward developing digital content.
- We need to encourage collaboration across the boundaries of all kinds of learning agencies, to break down the silos that separate not only library from museum from archives, but also that separate the institutions of formal learning from those that support informal learning.
- We need to foster the development of an ethic of constructive re-purposing of educational content and learning resources. Elements in the landscape that have been built for one specific purpose often have unanticipated uses in other learning contexts.
- We need to create systems that support customized learning experiences, tailored to the unique needs and interests of the individual learner. Such systems can capture, store, re-use and repurpose those unique experiences. And they can address not only on the needs of communities of place, but also on the requirements of communities of interest.
- In order to enhance access to the rich and diverse resources that support learning, we need to develop portals and recommender systems that enable the self-directed learner to identify, locate, evaluate and use resources that are relevant to their specific immediate needs.
- We need to assess carefully the implications of this new environment on library and information science (LIS) education. We are, I fear, continuing to prepare practitioners for a 19th century environment. I do *not* mean to say that we need to infuse the curriculum of LIS education with more technology—if anything, I think there is too much emphasis on technology

already. Instead, we need to equip practitioners to think broadly about the ever-changing environment in which they provide services, and teach them to focus their work of providing the services that the contemporary user needs, demands and values.

At IMLS, the National Leadership Grants program is one of our most important programs. Since 1996 when this program began, we have offered National Leadership Grants in three different categories to both museums and libraries, plus one category that encourages collaboration between museums and libraries. In the Library programs, those categories have been Preservation or Digitization, Continuing Education and Training; Research and Demonstration. In the Museum program, the categories have been Museums Online, Museums in the Community, Professional Practices. In addition, there is one category that spans both programs: Museum–Library Collaborations.

As our operations have continued to evolve over the past several years, and our interaction with the museum and library communities have progressed, we have come to realize that it is time for these programs to evolve as well. In exercising our responsibility to provide leadership for museums and libraries, next year we will be changing the categories for both museum and library grants.

Beginning with the 2005 grant cycle, IMLS will offer National Leadership Grants under three categories, the same in both Museum and the Library programs. Those categories will be

1. Advancing Learning Communities;
2. Building Digital Resources; and
3. Research and Demonstration.

A word about collaboration in the National Leadership Grant program: the elimination of a separate category for museum–library collaborations does not mean that the IMLS does not intend to continue to foster collaboration between and among museums and libraries. On the contrary, it signals our conviction that collaboration is such a central strategy that it should not be separated out as a single category, but rather integrated into all aspects of our programs. So, collaboration is encouraged in all three of the National Leadership Grants categories, and evaluation of proposals will be based in part upon a realistic incorporation of collaboration, where it is appropriate.

At IMLS, we provide leadership through our grant programs. Our National Leadership Grant program, in particular, has provided incentives to lead the field in a number of ways. We have funded a large number of digital projects in both museums and libraries. These important projects have dramatically enhanced public access to rich cultural heritage resources, available now for the first time to stimulate and support learners of all ages, in every circumstance. We envision the development of a seamless infrastructure for learning resources, blurring the boundaries between types of cultural heritage institutions, so that learners will not be inhibited by the traditional distinctions and practices of museums, libraries, archives and others agencies and professions.

In the 21st century environment of rapid change fostering an individual ethic and ability to learn throughout the lifetime is increasingly important, to maximize individual potential and social growth and stability.

Schools alone are not enough. Of course we continue to need schools, colleges and universities—we need the very best agencies of formal education that we can create. And libraries have always played an integral part in supporting curriculum and instruction in these agencies of formal education. Indeed, research demonstrates that the better supported and better integrated school and academic libraries are in the fabric of the institution, the better the parent institutions are able to achieve their goals.

But we need to go beyond our now-traditional notions of "education" and embrace a bold new vision of learning. We need to think beyond our institutional boxes and develop a seamless infrastructure for learning across all the social agencies and organizations that create, maintain, and provide access to resources that support learning.

Libraries—as well as museums, archives and other cultural agencies—are important elements in this web of learning. In fact, given our history of collaboration and our tradition of service, we can lead the way in demonstrating the potential and developing the reality. Our communities will demand it. If we do not provide it, someone else will.

References

De Rosa, C., Dempsey, L., and Willson, A. (2004). *The 2003 OCLC Environmental Scan: Pattern Recognition: A Report to the OCLC Membership*. OCLC Online Computer Library, Dublin, OH.

Falk, J. H., and Dierking, L. D. (2002). *Lessons Without Limit: How Free-Choice Learning is Transforming Education*. Altamira Press, New York.

Grimes, W. (2004). When the cashier is you. *New York Times* April 7, 2004.

Licata, P. G. (2004). Checking yourself out, quite literally. *New York Times* March 7, 2004.

Pink, D. (2001). School's out: get ready for the new age of individualized education. *Reason* October, 2001. Accessible online at: http://reason.com/0110/fe.dp.schools.shtml.

Zweizig, D. L. (1973). Predicting amount of library use: an empirical study of the role of the public library in the life of the adult public. Dissertation, Graduate School of Library Science, Syracuse University, Syracuse, NY.

The Evolving Relationships between Libraries and Scholarly Publishers: Metrics and Models

Craig Van Dyck and Christopher McKenzie
Scientific, Technical, and Medical Publishing, John Wiley & Sons, Inc., Hoboken, NJ, USA

I. Introduction

Libraries and professional publishers have long had a complicated relationship. This chapter seeks to explore some elements of this relationship in order to suggest new ways of regarding each of the parties and to generate an active dialog with the aim of improving collaboration and cooperation between them.

Libraries and professional publishers share a fundamental purpose: to provide information to seekers of it. Complicating this straightforward mandate are innumerable challenges. Libraries have limited budgets, and face a daunting amount of information from which to glean that content which will serve their unique user communities. Publishers serve multiple stakeholders: libraries, readers, authors, employees, professional society members, and shareholders. And for the professional publisher, the customer may mean different things. Is the customer the "end user"? Or, is the customer the entity (most often a library) that pays for the information the publisher provides? The answer varies depending on who is asking and who is asked. To editorial staff, the publisher may primarily serve its editors and authors; to sales or marketing staff the response is likely "whoever pays the bills."

Publishers also face intense competition for the limited material budget resources of libraries. This has never been more so than now when virtually all publishers have electronic offerings and the Internet has created a forum in which size matters, but less than it did at one time. Anyone with a few hundred dollars of software can readily post content on the Web (and many are doing so as evidenced by the proliferation of "blogs," for example), and claim to be a "publisher." This is, of course, a far cry from the credentialed information produced by professional publishers.

In this chapter, we will examine several models and metrics that have had a significant impact on publishing—for publishers and libraries—over the past 10 years. The theme common to all of these topics is that they have become significant over this period as a result of the dynamic and ever-expanding Web. First, we will explore how publishing decisions are made, and the role of the library in those decisions. Second, we will examine how information is marketed and sold to libraries and how this has changed in the e-environment. Third, we will look at how perceptions of content—on the part of the library and the publisher—have changed due to an important new metric: usage data. Finally, we focus on areas that are well suited for collaboration and cooperation between publishers and libraries. Within all of these topics, our goal will be to consider areas of mutual interest and the potential for collaboration and improved understanding.

II. How Publishers Decide What to Publish

A. Market Forces

1. Market Position

Wiley is a significant publisher with strengths in Science, Technical and Medical (STM) publishing, Professional/Trade publishing, and Higher Education publishing. Doing business since 1807, Wiley has evolved significantly, in order to persevere and flourish from the presidency of Jefferson to the present day. One of Wiley's biggest challenges is to continue to evolve as the marketplace evolves. Transforming its high-value journal content to an online format was an important recent challenge, and today's environment presents new challenges, such as tight library budgets, debates about new pricing models, and e-archiving. For a publisher (as for any company or organization), it is important to understand one's market position and one's strengths and weaknesses. Certain strategic questions then present themselves. Given our market position, are we positioned where we want to be, and if not, where do we want to be, and how can we get there from here? Given our strengths and weaknesses, are we capable to do the things we need to do to get from here to there, and if not, how can we add the missing capabilities, or should we alter our direction?

2. Living Among Market Forces

There are certain strong market forces at work today in STM journal publishing, including:

- Variable government funding of research, with some areas receiving healthy funding (for example, NIH research and development and basic research funding increased by 100% from 1996 to 2002; see AAAS, 2003, Trends in Federal R&D, FY 1990–2004, and Trends in Basic Research, FY 1976–2004).
- Weaker funding of libraries (for example, ARL libraries' expenditures increased by 33% from 1996 to 2002; see ARL, 2003, Expenditure Trends in All ARL Libraries, 1986–2002, Table 4).
- The formation of library buying consortia, e.g., OhioLINK, the Northeast Research Libraries (NERL), National Electronic Site Licensing Initiative (NESLI).
- Increasing development of new technology tools in aid of research, including the online journals that are now available at the scientists' desktop.
- The desire by end users for more and more online functionality, requiring ongoing investments to meet the requirements.
- New ideas about business models, such as Open Access, and participation by philanthropic institutions; for example, George Soros (http://www.soros.org/openaccess), Gordon and Betty Moore Foundation (http://www.moore.org/prgmareas_science.asp), Paul Allen (http://www.brainatlas.org/).

These forces create an environment with the following characteristics:

- Everyone in the value chain of scientific information is challenged to add value in new ways.
- Scientists enjoy the enhancement of research technology tools, and increasing integration of software and content.
- The parties who are trying to serve scientists—publishers, librarians, professional societies—struggle to provide all of the services that scientists demand, in an economic environment of scant increased capital.

Societies, publishers, and librarians share the common cause of facilitating scholarship.

B. Adding Value

In this environment, scientific publishers ask themselves, What value do we add? How can we add more value? What do we do that is indispensable, and no one can do better? Basically, publishers organize and fund the process by which scientific communities prepare and share their research results, and work together with scientists and librarians to innovate and implement to improve the scholarly communication process. This means:

- At conferences, editorial board meetings, campus and lab visits, and focus groups, speak with scientists and librarians about what they need in publications, resources, and tools.
- Based on these discussions, identify needed innovations, invest, and take risks in new initiatives to meet the expressed needs of scientists.
- Design, develop, produce, market, sell, distribute, and warehouse new and existing products.
- Recruit, gather, and pay scientists as editors-in-chief and editorial boards of journals and book series.
- Provide organization and funding for scientists as they conduct the peer review of scientific research articles.
- Provide online tools for authors and editors to prepare, submit, and review articles.
- Prepare manuscripts for print and online publication, and for long-term archiving: copyedit, XML tag, layout, proof, print and bind, present online, store in content management systems, distribute content to partners such as abstracting and indexing services and local hosts.
- Provide information about journals to libraries and authors.
- Distribute scientific content to libraries and individuals. With libraries, establish licenses for online access for their patrons.
- Pay the costs for all of the above.

Today, in addition to the publishers who have been providing these services for decades or centuries, there are new entrants such as the Scholarly Publishing and Academic Resources Coalition (SPARC; see http://www.arl.org/sparc) and the Public Library of Science (PLoS; http://www.publiclibraryofscience.org) who have also entered the field. Over time, the long-lived publishers have developed unique competencies:

Sustained focus. To survive, publishers must respond to new ideas that arise, and initiate new ideas themselves. Their survival depends on it. In response to new opportunities, publishers are continuously chasing the goal of "cheaper, faster, better." Over the years, there have been a number of initiatives that have arisen that intended to revolutionize scientific publishing. Some of these initiatives have come from the library community, but libraries have sometimes struggled to enlist the support of their university's faculty, who to some degree are already invested in the status quo, and do not necessarily see the library's problems as their own. Sometimes these initiatives have faded away as their proponents have moved on to other interests, or as the initiative has become sidetracked in disputes about standards (e.g., metadata), or superseded as the initiative morphs into something different (e.g., Scholar's Portal becomes institutional repositories and meta-searching). Publishers, while perhaps not the most nimble, have "outlasted the competition" by dint of their sustained focus.

Long-term development of new publishing models. Partly as a result of their sustained focus, and working closely with library, scientist, and technology partners, publishers have managed to develop new approaches that have evolved into robust models. Even before the Internet, the potential of electronic publishing was evident. Starting in the 1980s, publishers began to develop the Standard Generalized Markup Language (SGML). After a lengthy start-up period, SGML gained momentum when the US government's Department of Defense began to require that technical documentation be SGML-tagged. STM journal publishers used SGML to develop the MAJOUR (Modular Application for Journals) header Document Type Definition (DTD) in the late 1980s. In the early 1990s, publishers collaborated with libraries in four important electronic publishing experiments—the Chemical Online Retrieval Experiment (CORE; see Entlich *et al.*, 1997), The University Licensing Program (TULIP; see http://www.elsevier.com/wps/find/librariansinfo.librarians/tulip), Pricing Electronic Access to Knowledge (PEAK; see http://www.lib.umich.edu/retired/peak), and Red Sage (see Lucier and Brantley, 1995)—that shed light on how journals could be presented electronically in libraries. Around that time, Adobe Systems developed the Portable Document Format (PDF), which publishers quickly recognized as a breakthrough in the presentation of content online. SGML-tagged article headers (bibliographic information plus abstract and keywords) plus full-text PDF became a powerful package of electronic content. By the mid-1990s, with the advent of the World Wide Web, publishers and librarians developed innovative licensing models, including the "consortium" model, led by OhioLINK and Academic Press (University of Cincinnati, 1996–1997), which brought increased access to more content for more users. By the late 1990s, SGML had evolved into XML (Extensible Markup Language), offering better interoperability. And in the late 1990s, journal publishers created CrossRef to link the articles among the different publishers via reference links, in a "distributed aggregation" model, to use a term first applied to electronic publishing by Pieter Bolman of Academic Press while CrossRef was being created (private communication, 1998; Pentz, 2001). By this time, online journal publishing had settled into a few standard presentations and selling models.

Self-sustaining economic model. The subscription model has served scientific communication for centuries. In the online world, subscriptions and licenses continue to provide good value for libraries and users, and return on investment for publishers. There is a public debate about whether journal subscriptions are too expensive. Since the advent of online publishing, publishers have invested in increased functionality and volume of content, providing more value for money. Publishers continue to enhance the value of their offerings, especially by investing in technology, and also by continuing to improve their services, to continue to play a critical role in the scientific communication enterprise.

C. Profit and Loss, and "The Financials"

Publishers want to publish the work of the best scientists, in the areas where the most important and interesting science is being done, and where there is strong funding; for example, medicine tends to be such an area.

In today's economic environment, in the "hard" sciences where STM publishes, it is difficult to start up a new journal (absent a few million dollars of philanthropic start-up money, or a guaranteed subscriber base). This is a pity, because science continues "twigging," and research needs outlets. In previous years, the standard business model for a new journal was to lose money in the first 2 years, begin to make money in the third year, and have a cumulative profit by the fifth year. (In real life, it often took 7 years instead of 5.)

Today, in order to bring new content into journal publishing, it is more likely that a publisher will leverage an existing brand-name journal, and extend its reach into a new area, rather than starting up a small new journal that focuses only on that area.

The costs per published journal article have been estimated at around $4000. The Open Society Institute (2003, p. 16, Fig. B) gives $3750 per article for "editorial processing" which excludes print manufacturing; and King and Tenopir (1998, p. 9, Table 3) give $5000. Of course, journals' per-page costs differ based on variables such as print run, color vs. black and white, mathematics vs. straight text, page dimensions, and editorial office costs.

Revenue centers for publishers are:

- subscriptions/licenses: libraries, individuals;
- a share of member dues (for a society-owned journal);
- advertising;
- offprints and reprints of individual articles;
- sponsored supplements;
- color charges and page charges to authors (for some journals);
- pay-per-view for online articles;
- copyright fees for document delivery;
- license fees for digitized backfiles (old volumes of journals).

Of these revenue centers, subscriptions/licenses account for about 85% of revenue for most STM journals, though for medical journals with strong advertising the percentage of revenue comprised by subscriptions and licenses is more like 55%, with advertising at 35% (Credit Suisse First Boston, 2004, p. 27, Fig. 37).

Ultimately, decisions about what to publish are determined by the scientists themselves, in the persons of journal editors-in-chief, editorial boards, and peer reviewers. Scientific publishing is a communications loop within the community of scientists, who act as authors, editors, reviewers, and users, with publishers providing infrastructure, enabling technologies, capital, organization, and domain expertise in publishing.

Publishers look to librarians as an extremely important voice of the scientist/user. Working together with librarians, publishers know better how to reach the end users. In Section V we discuss specific areas where publishers and librarians can work together especially closely to improve the services that we jointly provide to scientific researchers.

III. How Publishers Market and Sell Content

A. Background

While almost all purchasing decisions are to some degree discretionary, professional (including scientific) publishers seek to publish "must have" content for practitioners in the fields in which they choose to publish. Examples readily come to mind from outside of the sciences: *Architectural Graphic Standards* (Wiley) is considered the design "bible" for practitioners; Jane's Information Group is famous for its defense and military books, magazines and reference works, including *Jane's Defense Weekly* and *Jane's Fighting Ships* (Woodbridge).

In STM publishing there is intense competition, but not the same type of competition as among producers of commodities, where price is often the most important differentiator. This is not the case among scientific publications. While there are some direct competitors in the field, these are few. An obvious example may be the *New England Journal of Medicine* and the *Journal of the American Medical Association*. Both are extremely prestigious and exclusive in terms of editorial content. Both however, can continue to be rigorous editorial selectors, maintaining their prestige because there is such an abundance of content published in medicine. Because of their import, then, while they may be competitors, selectors of content in the medical sciences will almost inevitably choose to buy both titles.

The competition among scientific publishers is a fight then for share of the limited budgets of libraries and other consumers of this specialized content. For STM publishers, libraries are an increasingly important market as the number of personal subscriptions held by scientists declines. One study, for example, shows that in a university setting in 1977, 60% of reading came from personal subscriptions and 25% from library subscriptions, but by 1990–1993, 36% of readings were from personal subscriptions and 54% from libraries (Tenopir and King, 2000, p. 30).

B. Marketing to Users and Acquirers

Prior to ubiquitous availability of electronic versions of journals, reference works, handbooks, encyclopedias and the other content types favored among scientific publishers, library purchasing decisions were made on a title-by-title basis. Bibliographers or subject specialists reviewed the content available in the areas for which they were responsible, then worked within the selection processes dictated by their organization either making the final decision about purchases or recommending selections to a committee or manager who had final budget and decision-making authority.

Orders for journals were placed through one or more subscription agents and payments to publishers were processed through agents who collected fees throughout the preceding year and then, less their discounts, passed these on to publishers. Until the accounts were tallied, publishers had no way of knowing how their journal renewal rates (or in the case of new titles, subscription rates) were faring. On the other side of the transaction, libraries making decisions at the title level could face a wide range of price increases even among the titles of one publisher and had no recourse except to choose not to renew a title if the price was found excessive relative to its value.

STM publishers did not employ direct sales representatives to broker their journal content, so marketers, working closely with their editorial colleagues, were organized within publishers much as bibliographers are organized within libraries: subject specialists managed the message that the publisher wanted to convey to a narrowly focused user and selector community.

Marketing staff remain engaged with the communities they are charged with serving. Marketing managers attend relevant scientific meetings and conferences. They often visit customers, arranging to speak with decision makers and "influencers." They work closely with editorial colleagues to manage the needs of editorial boards and staff. In this capacity, they also act as a valuable conduit of information *from* the market *to* the publishing house. Journal editors are also authors and users of STM content; publishers rely on them to act as representative voices for their fellow researchers. This constant communication "loop" allows publishers to maintain currency, and therefore to adapt better to the changes in the fields in which they publish.

Managing a limited marketing budget, these subject-oriented marketing specialists face the challenge of making sure that the message that they are communicating is received by the appropriate person(s). In almost all cases selection decisions are made by more than one person and the message that publishers need to convey may take several forms, customized to the particular recipients.

The role of the librarian also needs to be considered within a geographic or cultural context. Librarians in parts of Asia, for instance, often have a different role in decision-making authority relative to the faculties they support compared to their western counterparts. The faculty are more closely involved in the buying decision and often manage the budget that funds such purchases. While the librarians cannot be ignored, the faculty role may be much more direct and influential than in North America or Europe in purchasing decisions. In other regions, subscription agents may play a much bigger role than in the US or Western Europe. In Mexico, publicly funded libraries must bid on content every year, a highly competitive process managed by agents. Publishers in these cases may be quite divorced from the decision maker, and therefore must keep the information brokers involved and informed.

Complicating all of this is the vast proliferation of electronic, web-based offerings of publishers. All major and most minor STM publishers, including societies, have directly or indirectly made their content available via the Web. This change, affecting all stages of the publishing process, has led to major changes in marketing and selling content to information seekers. And of course, libraries have had to adapt as well, now having to meet the needs of the so-called "virtual patron."

Up to this point, patrons had to travel to a physical place—generally the library—where the physical artifacts were housed and archived. Often the process of finding and retrieving the information necessitated the involvement of a librarian. The user of library-funded resources may continue to use the finding aids provided by the library, but is now likely to be able to access these resources from his or her desk rather than going to a separate repository. This has had profound impact on library usage and the ways in which libraries serve their customers, just as electronic publishing has had a similarly profound impact on the way publishers serve their customers: the self-same authors and libraries.

C. New Models of Selling Content

The most fundamental change in the commercial relationship between libraries and publishers has been the shift away from the "terminal transaction." That is, when only purchasing print, the buyer paid a fee (in most cases, the "list price") and received the product (journal/book/whatever) and while publishers sought to maintain an ongoing relationship with the libraries that bought their non-monographic products, there was no requirement that they do so. In the Web environment, however, the relationship between the library (assuming for this purpose, the library is the consumer) and the publisher is much more iterative. Publishers have moved from offering content for purchase in a one-time transaction to an ongoing *licensing* relationship, in which the library pays fees regularly to maintain access to the content offered via the publisher or intermediary's website. There are, of course, variations on this involving, for example, archival rights, perpetual access, print vs. electronic-only, but the rights and responsibilities inherent in this kind of transaction are markedly different from the sale of print products.

At the advent of this new period in publishing, publishers and libraries found that they lacked staff familiar with electronic publishing. Acting to change this, library schools and practicing librarians began to focus on electronic offerings. Libraries created new positions such as "Electronic Resource Officers." Publishers did the same, adding technical staff to manage their electronic services, educating editorial and production staff in new methods of publishing electronic content, and adding marketing and sales staff to negotiate usage and pricing terms with customers.

Offering content in new ways meant that users were no longer bound by the same restrictions of time and location that they had been when content was only available in print at the library. While this was recognized as a new and exciting era for publishers, it also caused concern about protecting the validity and provenance of intellectual property in the new electronic environment; publishers wanted to continue to effectively disseminate and protect the content. Publishers reacted to this by imposing restrictions on the use of electronic content. These limitations included requiring user names and passwords for anyone wanting access to licensed content. Others placed strict geographic limitations (one publisher authenticated users by Internet Protocol (IP) address but tried to ensure that access was only within one building). Often these restrictions conflicted with both the intentions and the technological infrastructure of libraries. University libraries, for instance, often could not distinguish their IP ranges from those of an academic department. User names and passwords for all but the most arcane material simply could not be administered effectively in a large academic or global corporate environment.

Since that time, publishers found more moderate methods, such as IP or proxy address authentication, of providing access to content, so that users have the access they need to content when they need it, without unnecessary obstructions, but risks remain (the recent experience of the music recording industry gives pause to owners and distributors of intellectual property). This added functionality has also led users to have new requirements and expectations, to which publishers and librarians have to respond with astonishing speed. Users have quickly grown dependent, for example, on linking arrangements between primary and secondary resources, and expect that these resources will be available to them 24 hours a day, 7 days a week from their desktops.

Publishers realized that in the transition from a print-based business to new electronic focused licensing arrangements, new pricing metrics would have to be developed and tested. Tenopir and King (2000, p. 44) state it clearly: "Pricing will be the most important issue that publishers, libraries, and scientists will face over the next decade." Questions quickly arose about how to ensure that the business that publishers already had could be preserved, while presumably the increased utility and extended "reach" of the electronic version of content would increase its value for customers and users. Publishers also made significant investments in technology to make their content available electronically, and it was expected that these investments would be recouped in increased sales. Libraries, however, often assume that if publishers no longer have to produce as much print, costs of the same content available electronically will remain neutral or even decline. Tenopir and King (2000, p. 372, citing Odlyzko, Boyce and others) cite at least two studies that conclude that the costs associated with publishing electronic journals "do not decrease appreciably" relative to print publishing costs. In most cases, this tension has led to a conservative response. Publishers have often based the pricing for electronic content on a customer's historical print spending. Long-term licenses (those for more than a single year) can also offer customers more favorable terms.

At the same time, the role of library consortia has expanded rapidly. Academic libraries turned to the consortium to which they belonged to manage the process of negotiating licenses for newly available content. Today, consortia often represent a large proportion of STM publishers' customers. The buying and negotiating clout of these consortia led publishers to consider new ways to maintain and grow their business with the consortium's members. Publishers offer incentives to consortium members to secure licensing agreements; typically, this is access to content to which the library would not have access if it acted alone. For this access to new content, the library paid little or no cost beyond its own subscription-based fees. For instance, if two libraries, A and B, are members of a consortium and each has subscriptions from publisher X (A has 20 and B has 30 titles), the publisher might offer access to both the titles to which the library subscribes but also to any titles subscribed to by the other consortium member. In this example,

A would gain access to at least 10 new titles to which it had previously had no subscription. B would gain access to any of the 20 in A's collection to which it did not already have a subscription.

Publishers embraced this idea because the opportunity cost is low, and the value it delivers to customers is substantial. Growth in the journal business generally comes from taking market share, not from significant growth in library subscriptions. Consortium arrangements generally included proscriptions against cancellations during the license term, so the publisher secured its revenue from that group of customers for the license duration. Similarly, libraries have seen these arrangements as a means of significantly augmenting their collections at modest cost. Instead of having to cut titles, they could, with one license agreement, potentially add literally hundreds of titles to their electronic offerings.

These agreements proliferated in the 1990s involving most library consortia and key STM (and non-STM) publishers. These agreements also mutated. Some consortia were so large that in aggregate its members subscribed to virtually all of a publisher's titles, and it made sense then to offer the consortium all of a publisher's titles as the key benefit of the consortium license. Or, more simply, the consortium had such significant negotiating authority that it could simply demand this benefit, and publishers often acquiesced.

These types of licenses, characterized by Ken Frazier (2001), University of Wisconsin—Madison Library Director, as "the Big Deal," have become the subject of much debate among publishers and libraries as both have had time to consider the consequences of these arrangements. Big Deal proponents cite the fact that users no longer face the limitation of having access only to the titles to which the institution subscribed in prior years. For a smaller institution, the Big Deal can expand its scientific journal collection in unprecedented ways. Usage data also bear out the maxim that if content is made available, users will use it. OhioLINK data illustrate this point: "Between April 2000 and March 2001, of the 1,306,000 articles downloaded, 58% were from journals not held in print at the downloading patron's library. For small colleges, 90–95% of articles downloaded were from newly accessible electronic journals in 2000" (Tenopir, 2003, p. 19). Publishers like the Big Deal because it acts as a powerful incentive for libraries to sign agreements that secure their revenue for a fixed period of time.

Critics assert that the Big Deal encourages publishers to continue to publish content that is of questionable value; in essence, that these arrangements allow journals that would otherwise "die" to continue to live because they are lumped in with the valuable content the library and its users really need. The case of either side can be bolstered by usage data. Proponents of the Big Deal cite usage information which shows that if you make it available, users will access it—and for this access, the institution has paid a fraction of the list prices of the accessed journals. Opponents suggest that this means simply that some use is inevitable, but that if users did not have access to this unsubscribed content, there would be no serious drawbacks; instead, this casual use creates a false perception that these titles have value that would quickly evaporate if left to actual subscription variations.

These arrangements serve users well, especially at smaller institutions that may have had access to only limited journal resources prior to participating in a consortium's Big Deal. Publishers like to have their content available to the largest possible user community, because it then gets cited more often, potentially raising their journals' Impact Factors (see http://www.isinet.com/essays/journalcitationreports/7.html). Authors like the increased exposure of their articles. In this way, the same content has more value to users simply by virtue of a new pricing and distribution model.

Some of these agreements have now been in place for between 5 and 7 years. As the current licenses are being renegotiated between major consortia and publishers, both are considering the question of whether the Big Deal should continue to be available as a standard pricing model, or whether it needs to be modified or discarded.

Libraries consider this question within an environment of severe budget constraints. Libraries are reviewing the merits of the historical basis of the pricing—and with detailed usage information, are trying to pay less for little used content. Publishers, using the same data (which generally show rapid adoption of the use of their electronic content), assert that these agreements offer significant value for libraries.

It seems unlikely that publishers or libraries will abandon the Big Deal (or its variants) anytime soon, but both are looking for new models that will serve the needs of customers for whom this model is not a good fit. For those customers outside of consortia or for whom the Big Deal is unattractive or for smaller customers, publishers have typically offered alternative options. However, these models may have limits on what the customer may access, or other usage and access restrictions. Generally, these electronic offerings are also made at the title level. The bottom line seems to be that the Big Deal is a good deal for libraries (and their users) that can afford it, but publishers must offer alternatives for libraries with fewer resources or greater specialization.

D. Technology and New Sales Models

Publishers are also trying to adapt pricing models to different "units" of content; some have referred to this as the "article economy." In these scenarios, the customer may not be the library, but rather the "end user."

The user pays "by-the-drink" for the article(s) he or she downloads. In these arrangements, for pricing purposes, it rarely matters which title the article is from, the publisher charges a flat rate per article, whether the journal itself is a relatively high-cost biomedical title or an inexpensive business title. This is certainly easier to administer for publishers and customers, but it fails to account for the perceived relative value of the content. Paying $25, for instance, for an article from a specialized medical journal when it helps to lead a pharmaceutical firm to a multi-million dollar drug discovery would be a bargain. The same article, on the other hand, could be merely ancillary "nice-to-have" reading for other users. Publishers, scientists, and librarians should continue to work together to define more sophisticated measures for value and effectiveness, and to consider together how these measures could be incorporated into new sales models at the article level.

Technology also facilitates the development of new sales models. Present Web services allow use for fixed periods of time, for example. Perhaps libraries will soon pay a fixed fee to have access to selected or all content a publisher offers for a limited period of time (a semester, for instance). Publishers may also offer more granular elements of content: the paragraph, page, or figure.

Some library customers, particularly those outside of academia, have been willing to forego the subscription model entirely, relying instead on delivery of content on an "as needed" basis. Libraries that rely on this "just-in-time" method of securing the content only when sought by a user expect that their needs will be satisfied by document delivery suppliers, interlibrary loan, or publishers themselves via "pay-per-view" functionality. Similarly, some aggregators hope to function as the primary gateway to content from many publishers, figuring that their robust bibliographic records can offer a "one-stop" option, seamlessly linking to publishers' sites to access full text.

These issues also have implications for content other than journals, as well as newly available electronic versions of previously published content, e.g., backfiles (digitized versions of old journal volumes not previously available electronically).

Increasingly, publishers are finding that libraries are very interested in backfiles of quality journals. Content in the physical sciences has a much longer "shelf life" than in medicine, and libraries and users want to be able to integrate that older content with the newer content that is available online. Alternatively, content in computer science may have little value (except to scientific historians) shortly after publication. And as libraries shift their resourcing away from print and towards digital, they want to save shelf space by replacing older print copies with their digital versions.

In addition to different product offerings, publishers need to deal differently with different customer types. Corporate customers sometimes have a different mandate than their academic counterparts in that their users may be more time-sensitive than cost-sensitive. These libraries generally focus on immediate access to content and are less concerned with the role of preservation or storage of information for future scholarship. Publishers have had to respond to these demands, and more are doing so by offering alternatives to standard pricing and subscription models and new access methods, both enabled by technological innovations. These technologies allow publishers to deal directly with the end user as consumer. Pay-per-view and other "point of sale" purchases are often available for users who are not affiliated with an institutional customer; this increases readership of publishers' content. Examples include pricing based on usage and alternative access options, some of which are discussed below.

IV. Usage Data as a Metric

A. Usage Data: What It Does and Does Not Show

One of the great advantages of electronic publishing has been the facility of collecting usage data. No longer dependent on anecdotal or unreliable methods for trying to determine the frequency of use of journals (e.g., counting unshelved print copies of a title), librarians and publishers have been given the ability to determine precisely how frequently, when, where and in some cases, by whom, licensed electronic content is used. Further, it is possible, often, to determine through what source a user came to the publisher's content, whether from browsing or searching the publisher's own service or through an abstracting and indexing (A&I) service or a CrossRef (see below) link within a reference.

It remains difficult to answer many questions about use with currently collected data, including *why* a user selected certain material, or what he or she gained, if anything, from it. Beyond "value," how can we measure *effectiveness*? Despite these persistent questions, usage data can inform both economic and editorial decisions for both publishers and librarians.

A caveat is worth noting. Usage of licensed electronic materials has grown dramatically over a very short period. This rate of growth will be inevitably slower as the number of libraries providing access and the number of titles available electronically stabilizes. Given this, one example illustrates the need for caution in trying to interpret usage statistics: from 1998 to 2001, OhioLINK growth of use of electronic journals (in full-text downloads) increased by 464%; during the same time period, average serial costs (across disciplines) increased by 23.1% (ARL, 2001–2002). Thus, one could conclude that relative to use, electronic journals provide good value for money. Conversely, OhioLINK statistics show that during a 1-year period beginning in April 1999, 85% of use of electronic journals in OhioLINK's collection came from 40% of the available titles (Tenopir, 2003, p. 19). The lesser-used titles might be targeted first if subscription cancellations are necessary.

Usage data tell publishers how users access information given a variety of access options. At the same time, all usage is not the same. One download is not necessarily the same as another download. Utility to the user cannot be measured with available statistics, so usage must be considered within the context of other conditions. Usage is also an artifact. It tells only what has come before; it provides no insight into the future. As research interests tend to change frequently, those trying to make decisions (whether editorial or economic) using historical usage data must also consider other salient environmental data specific to their institution and user community.

B. Usage Data and Publishing Decisions

Usage trends can and do inform publishing decisions. Trends in usage indicate subject areas in which research is apparently headed, and this can lead publishers to anticipate growing demand. Publishers can use these data as a factor in deciding subject areas in which to expand their offerings: as a result, we have seen additional publications in fields such as proteomics, genomics, bioinformatics, and fuel cells. In the past 3 years, although it is a difficult economic environment for starting new journals, Wiley has launched journals with titles such as these: *Chemistry and Biodiversity, Comparative and Functional Genomics, Engineering in the Life Sciences, Fuel Cells, Functional Genomics, Gene Function and Disease, Journal of Genetic Medicine,* and *Proteomics* (see http://www.interscience.wiley.com).

In medicine and the life sciences, usage data inform decisions about what areas to speak with scientists about, and potentially point to journal content to acquire and what functionality should be supported with additional investment. For example, usage data show that a high proportion of users "land" on full-text articles as a result of searches in abstracting and indexing sources. Clearly, abstracts have value, and with the concurrent emergence of handheld technologies, it seems apparent that abstracts delivered to the palm would be popular. Physicians were rapid adopters of handheld technologies, and this has led some publishers to choose medical titles as the first titles to offer downloadable abstracts for use on handheld devices. For example, Wiley currently offers the tables of contents and abstracts of articles in 23 medical journals delivered to Personal Digital Assistants (PDAs); see http://www3.interscience.wiley.com/aboutus/mobileedition.html.

C. Usage Data and Pricing

Usage data are emerging as an important aspect of discussions about the pricing of electronic information. Librarians, armed with title-by-title costs of journals and other electronic content, are scrutinizing their collections in new ways, asking tough questions of those information providers whose costs are perceived to be above the norm. Publishers too are using these data to support the assertion that electronic content has been widely adopted by users, and that this should encourage renewed long-term commitments to journal packages.

Alternatively, usage data can precipitate new pricing models. These have so far been of two types: pure usage, and a hybrid of usage and subscriptions.

"Pure" usage-based models can work in various ways. Perhaps the most extreme in terms of its difference from currently available pricing schema would have no predetermined conditions. Access to content would be purely discretionary and consumers would pay "as they go." In this scenario, consumers (whether institution or individual) would have no obligation to any set amount of consumption. They would decide on an ad hoc basis when to select content for which they would pay a set fee which could be the same for all of a publisher's titles, or could vary depending on the title or medium, for example, book chapters could be available at a different rate per download than journal articles.

A variation on this model is that the consumer could be offered favorable pricing terms in exchange for an up-front commitment to a certain amount of downloading. In exchange, publishers would be willing to offer lower per unit prices. This reduces costs, but also flexibility to control spending. Some databases are priced this way.

A hybrid approach may be the best solution. For example, a library could pay a set subscription-based fee for access to some titles and at the same time buy access to unsubscribed content on an as-needed basis. Striking the right balance between these two pricing models is a challenge facing collection development librarians and publishers.

The 80/20 rule may apply to the use of scientific and other professional content, but it would be antithetical to the good of science to suggest that publishers should only publish the 20% of currently available journals that are most heavily used. No matter how reasonable a pricing model may be, purely economic analysis does not necessarily work when considered alongside the mission of STM publishers and librarians: the dissemination of important scholarship. By making the initial investments in start-up journals that may or may not succeed, publishers take a risk because they consider the content being produced important. Concurrently, librarians pay more, relatively, for less critical journals. Publishers and libraries value the good of science more than strict economic rationality to ensure that highly specialized, little-used research material continues to be published and collected.

The argument against the Big Deal is that the deal protects "lesser-quality" journals. This goes to the heart of the central questions in collection development. Who is to determine what content is worth supporting, and should librarians give users what they want, or what the librarian thinks they need?

Librarians and publishers can work together to ensure that the content being published, and the methods used to acquire access to it, evolve to meet the dynamic needs of the users we both serve.

V. Areas of Potential Cooperation and Collaboration

Librarians and publishers share common cause. They share the mission of aiding scientists, they share the same economic conditions, and they share the imperative to demonstrate that they continue to add significant value in the modern, online world. There is a great deal of overlapping mutual interest. Both parties would benefit from increased efficiencies in the system (standards), increased integration of technology tools, and increased funding for libraries. There are important ways in which these shared goals can be leveraged for collaboration on mutually beneficial initiatives.

A. Standards

All such initiatives will benefit from agreed-upon industry standards. Librarians and publishers have worked together extensively on standards for many years; although, in truth both parties can do more to start their standards activities together, rather than starting them within their own community and only later trying to enlist the other.

Examples of important joint standards activities are the following ANSI/NISO standards: OpenURL (Z39.88), Digital Object Identifier (DOI) syntax (Z39.84), Serial Item and Contribution Identifier (Z39.56), Bibliographic References (Z39.29), Information Retrieval (Z39.50), and Dublin Core (Z39.85). Other joint standards activities include International Standard Serial Number (ISSN), International Standard Book Number (ISBN), electronic-archiving (Digital Library Federation, 2003), usage statistics (COUNTER; see http://www.projectcounter.org/), and reference linking.

But some questions entail significant business issues, and do not lend themselves to the standards process, but still need cooperative collaboration among librarians and publishers.

B. New Pricing Models

Many librarians have been frustrated with the subscription model that is based on historical holdings for some years. Many libraries have needed to reduce their collections by canceling subscriptions to journals that they perceive as only marginally adding value for their users.

With the advent of online publishing, the "Big Deal" model (with restrictions on subscription cancellations, in return for access to more journals for more users, and caps on price increases for the term of the license) has achieved significant uptake. This model was pioneered by libraries, particularly OhioLINK, the consortium of libraries in the state of Ohio. However, by accepting the Big Deal, librarians feel that they are "locked in" to levels of spending and to subscriptions, and they want to reassert more control over the collections (Frazier, 2001).

"Usage-based pricing" is an alternative model that has been much discussed in the past few years. In this model, libraries would pay based on the articles that their patrons actually use, rather than based on subscriptions to journals whose articles might or might not be used by the library's patrons. However, so far there is not an accepted usage-based model with which librarians are comfortable; the amount that the library might be charged is too unpredictable. And meanwhile models such as the Big Deal have increased usage dramatically, and driven down the per-use cost that large libraries pay via subscriptions (Stange, 2003, slide 12). Wiley and other publishers are currently working closely with libraries to find a mutually agreeable method to add usage as a metric to the cost/value equation.

Pricing is not an area that lends itself to formal standardization. However, market forces have a way of evolving towards de facto standardized forms. Publishers and librarians will continue to experiment with new pricing models, and eventually the best approaches will survive. In the search for new and sustainable business models, controlling the risks of experiments is key. In pricing experiments, both the publishers and libraries will want protection from unforeseen financial consequences. For the next few years, such risk-controlled experiments will shed important light on the results of different pricing schemes. But it will likely take several more years before experiments can be concluded, the results digested, and new models be added or replace the current ones.

C. Archiving Digital Content

Libraries have been de facto redundant archives of print copies. In the electronic world, in contrast, libraries normally access content from publishers' servers. Who is responsible to ensure the long-term preservation of the electronic resources that scientists use as part of their research? If libraries have carried that responsibility in the print world, should libraries continue with the responsibility in the electronic world even though libraries are not hosting the electronic content?

Or should publishers take on the responsibility? Libraries have expressed reservations that publishers cannot be relied upon to ensure long-term preservation of the e-content: What if the publisher goes out of business, or, if the content ceases to be a financial asset, will publishers be willing to continue to bear the costs of preserving the content? Publishers have tried to reassure librarians that they will indeed preserve content, but librarians have not responded positively to what is essentially a "Trust us" argument by publishers.

As a result, publishers have agreed with librarians that long-term preservation of electronic content is important, and that solutions need to be found. At present there are a number of electronic archive projects underway that have been initiated by libraries and that include publishers' participation: JSTOR (http://www.jstor.org/about/earchive.html) and LOCKSS (http://www.lockss.org/), both with Mellon funding; the California Digital Library (http://www.cdlib.org/programs/digital_preservation.html); the British Library (http://www.bl.uk); and the Koninklijke Bibliotheek, the National Library of the Netherlands (http://www.kb.nl/).

There are a number of issues yet to be solved:

- Standard formats and metadata packaging for sending content to the archive.
- Agreement on what must be archived, vs. what is optional (for example, must non-article items such as meeting announcements and advertisements be e-archived? What about user interface functionality such as links, search options, and personalization?).
- How to "future-proof" the data so that today's formats will be accessible and readable by users in the future?
- How to handle non-standard formats that often occur as supplementary material, e.g., audio and video files?
- What restrictions are there on access to the archived content?
- At what point should there be no restrictions on access?
- Who pays the costs of the e-archive?
- Can the archive exploit the archived content in ways to recover costs, i.e., to make money?
- Who "owns" the archive, and the content in it?
- Who has governance over the policies and practices of the archive?

These are difficult questions. But libraries and publishers are on the right track. Working together, going step by step, these questions will be answered.

The Harvard University Library provides a good summary of electronic journal archiving issues (Digital Library Federation, 2003).

Over the past 15 years, one thing has become clear—what seems intractable today, will be resolved tomorrow. Stakeholders in electronic publishing have demonstrated their ability to work together to continue the evolution of electronic publishing. There is more than enough forward momentum to overcome obstacles. The more powerful user functionality of electronic publishing makes electronic content the preferred medium for scientists, so those serving the scientists (librarians and publishers) are strongly motivated to solve the problems that arise.

D. New Publishing Models

The journal is the package in which scientific research is disseminated; journal articles are the components of the package. Journals present a collection of articles that are within the journal's defined scope and focus, and that are published with the imprimatur of the journals' Editorial Boards, who manage the all-important peer review process. This model has existed for hundreds of years, and has been a key contributor to the advancement of science. The system has provided a validation process to help scientists sift through the overload of new information, including metrics such as the Impact Factor to provide at least some degree of quantitative measurement of value. Libraries have been the gateway to this content: organizing, preserving, and helping users get to it. New publishing models challenge not only the scientific communication method currently employed by publishers and authors, but also libraries which must be prepared to evolve if access to information is to remain efficient.

Electronic publishing presents new possibilities. Some might ask whether it is necessary to subscribe to an entire journal, when online searching tools can bring you to the precise articles that you want, no matter what journals they are in, and you can purchase access one article at a time. They ask whether a journal's formal peer review process is necessary, when authors can post their own articles, or send them to servers that collect articles in a particular discipline (e.g., the physics preprint server arXiv; see http://www.arxiv.org) and anyone in the community can read and openly comment on these articles. Can technology horsepower and tools handle all of the tasks that journals and publishers perform, at less cost? Can libraries manage to find the sources their users demand when users are dispersed so broadly? Will these sources be preserved, with historical access available?

Publishers must be prepared to modify their own models, and to compete with new models; and may the best models and the best service providers win.

That is the open market of ideas, models, and competition. When we think of potential alternative models for the dissemination of scientific research, there are several possibilities visible on the horizon. For each of these, there is the opportunity for publishers and libraries to work through the issues together.

1. Self-publishing

Authors can post their own articles. By adhering to the Open Archives Initiative Protocol for Metadata Harvesting (http://www.openarchives.org/OAI/openarchivesprotocol.html), the articles are discoverable via their metadata. The full text can be indexed by Google and other search engines. By using the DOI (http://www.doi.org), authors can ensure that their articles are persistently available on the Web. By participating in CrossRef (http://www.crossref.org), authors can take advantage of reference linking, forward linking, and other features such as cross-publisher full-text searching. One can imagine standards being established so that a loose federation (e.g., of scientific societies) could organize all of these self-published articles. How would peer review, or some other form of validation, work? Either there could be zero peer review, or some new form of organized online peer review could be established, perhaps akin to the readers' reviews in Amazon.com. We do not know yet whether authors will be willing to let go of the validation that comes by being associated with a respected journal.

2. Institutional Repositories

Authors can look to their institutions to post their articles. As Crow (2002) describes well, many universities are working to organize the intellectual output of their populations, and to make it available on the Web across universities. Whether faculty will be willing to adhere to standards set by their universities and to give over the management of their articles to the school, and whether universities will be able to support the costs of institutional repositories, and whether commercial firms will participate in such a network are not clear.

3. Open Access

New online journals have started up that are available for free to users. Authors pay a fee to publish. This is a conventional journal publishing model, but with the economic burden shifted from the readers (or libraries, as the readers' proxies) to the authors (who in many cases are the same people as the readers) or their proxies, the institutions that employ them. Initially, the author fees were in the $500–1500 range (BioMedCentral and Public Library of Science, respectively). However, as stated above (see Section II.C), the per-article publishing costs exceed that amount, and it is not clear whether authors will support this model, so the financial sustainability of Open Access publishing is not yet confirmed. Public Library of Science and BioMedCentral are leading Open Access publishers (http://www.publiclibraryofscience.org and http://www.biomedcentral.com) to explore this approach. By mid 2004, author fees from $3000 (Springer's "Open Choice") to $6000 (American Society of Human Genetics) had been announced. Meanwhile, government and private research funders had begun to ask for Open Access to articles they funded.

In conclusion, these different future models are not mutually exclusive. We will likely see a mixture of models living together, serving different situations and needs.

E. Distributed Aggregation: CrossRef

One of the weaknesses in the present system is that each publisher is a silo of content, whereas users identify with journals, not with publishers. CrossRef is an attempt to bridge the silos via reference linking. CrossRef is a non-profit member organization including 300 scholarly publishers; see http://www.crossref.org. CrossRef was launched by a small group of leading journal publishers with the purpose of establishing a mechanism to link from references to the cited articles. This effort has been very successful. Using the DOI, CrossRef publishers now make millions of links each year, which has added an important new functionality to online journals (see http://www.crossref.org/01company/00introduction.html and click on Annual Report, letter from the executive director and the chairman).

CrossRef's statistics as of early 2004 show that there are 10.5 million articles and 9500 journals in this metadata database; about 20 million online links are embedded in the publishers' content per year, with over 5 million real-time end-user links ("DOI resolutions") per month (CrossRef, 2004).

Based on this success, CrossRef now plans to add further functionality such as forward linking, and CrossRef is discussing with librarians and scientists what further functionalities it could be helpful for CrossRef to provide.

Having been developed by publishers, CrossRef initially met with some questions from librarians. This is a good example of how working together earlier could have reduced the potential for misunderstanding. However, 4 years later, CrossRef is now being asked by some librarians to broaden its scope much more widely. In meetings such as CrossRef's Library Advisory Board, as well as in publishers' own library advisory boards, librarians look to CrossRef as the logical entity to eliminate barriers and to improve users' experience.

For example, each publisher's website is different, with different interfaces, different log-ins, different navigation, and different transactional protocols. All of these differences detract from the users' experience. Librarians ask, why cannot publishers establish one interface? Would not CrossRef be the perfect forum?

Publishers are conflicted about this. Publishers view their online publishing engines as sources of competitive advantage. And publishers want to be differentiated from their competitors. On the other hand, there are undeniably inefficiencies for the users. How to reconcile these differences?

As usual, going step by step is the best way forward. Having achieved cross-publisher reference linking and forward linking, CrossRef is beginning to seriously consider further steps. Librarians are asking for full-text searching across all CrossRef member publishers' content, "one-stop shop" for interlibrary loan and document delivery, cross-platform authentication of users, more integration of discovery and other tools and content, and fewer barriers to access. Publishers are viewing these requests as opportunities to enhance their services to libraries and users.

VI. Conclusions

In the age of online publishing, collaboration is a must-have, not a nice-to-have condition. No one can figure it all out on their own. All stakeholders must work together to understand the best ways to harness the amazing potential of new technologies. It is unfortunate that the natural division between seller and buyer has interrupted publishers' and librarians' natural shared commitment to improving scholarly communication. Fortunately, all parties show the ability to work together where there is potential to innovate, even while in parallel continuing to discuss the issues that require further reconciliation. Librarians and publishers share too much common cause to fail to overcome any obstacles to making the most of the opportunities that the 21st century offers.

Acknowledgements

The authors express their thanks to Nancy Roderer for advice and Nicole Luce Rizzo for help with research for this chapter.

References

AAAS (2003). American Association for the Advancement of Science Guide to R&D Funding Data, http://www.aaas.org/spp/rd/guihist.htm.
ARL (2001–2002). Association of Research Libraries Statistics, http://www.arl.org/stats/arlstat/02pub/intro02.html.
ARL (2003). Association of Research Libraries Statistics, http://www.arl.org/stats/arlstat/index.html.
Credit Suisse First Boston: Equity Research: Europe: Industry: Scientific, Technical, and Medical Publishing: Three Pillars of Wisdom. Simon Mays-Smith, Giasone Salati, Neil Shelton, Nick Bertolotti, Frederik Kooij. STM Publishing 070804 (April 6, 2004).
CrossRef (2004). Newsletter, March 2004, doi:10.5555/monthly_newsletter, http://dx.doi.org/10.5555/monthly_newsletter.
Crow, R. (2002). The case for Institutional Repositories: A SPARC position paper, http://www.arl.org/sparc/IR/ir.html.
Digital Library Federation (2003). Archiving electronic journals: research funded by the Andrew W. Mellon Foundation, http://www.diglib.org/preserve/ejp.htm.
Entlich, R., Olsen, J., Garson, L., Lesk, M., Normore, L., and Weibel, S. (1997). Making a digital library: the content of the CORE project. *ACM Transactions on Information Systems* 15(2), 103–123. doi:10.1145/248625.248627, http://dx.doi.org/10.1145/248625.248627.
Frazier, K. (2001). The librarians' dilemma: contemplating the costs of the "Big Deal". *D-Lib Magazine* 7(3). March 2001, doi:10.1045/march2001-frazier, http://dx.doi.org/10.1045/march2001-frazier.
King, D., and Tenopir, C. (1998). Economic cost models of scientific scholarly journals. Paper presented to the *ICSU Workshop*, Oxford, UK, March–April 1998, http://www.bodley.ox.ac.uk/icsu/kingppr.htm.
Lucier, R., and Brantley, P. (1995). The Red Sage project: an experimental digital journal library for the health sciences. *D-Lib Magazine* August, doi:10.1045/august95-lucier-brantley, http://dx.doi.org/10.1045/august95-lucier-brantley.
Open Society Institute (July 2003). Guide to launching an Open Access journal, http://www.soros.org/openaccess/oajguides/business_planning.pdf.
Pentz, E. (2001). CrossRef: a collaborative linking network. *Issues in Science and Technology Librarianship* no. 29, Winter 2001, http://www.library.ucsb.edu/istl/01-winter/article1.html.
Stange, K. (2003). Complexity and costs in the purchasing process. PowerPoint presentation delivered at the Association of Subscription Agents and Intermediaries Conference, 24–25 February 2003, London, http://www.subscription-agents.org/conference/200302/index.html.
Tenopir, C. (2003). *Use and Users of Electronic Library Resources: An Overview and Analysis of Recent Research Studies*. Council on Library and Information Resources, Washington, DC, pub120, http://www.clir.org/pubs/abstract/pub120abst.html.
Tenopir, C., and King, D. (2000). *Towards Electronic Journals: Realities for Scientists, Librarians, and Publishers*. Special Libraries Association, Washington, DC, http://www.sla.org/content/Shop/Resources/titlelist/towelecjnl.cfm.
University of Cincinnati (1996–1997). Ohio academic libraries purchase electronic rights to Academic Press journals, Winter 1996–1997 Technical Update, http://www.uc.edu/ucitnow/winter_97/lib.html.

The US Government and E-Government: Two Steps Forward, One Step Backwards?

Peter Hernon[a] and Robert E. Dugan[b]
[a]Graduate School of Library and Information Science, Simmons College, 300 The Fenway, Boston, MA 02115-5898, USA
[b]Sawyer Library, Suffolk University, 8 Ashburton Place, Boston, MA 02108, USA

During the 1990s, electronic government, commonly known as e-gov, materialized "as a dynamic concept," but one having "varying meaning and significance" (Relyea, 2003, p. 379). Various policy instruments have shaped this concept and its application. Such instruments "seek to promote the use of new IT [information technology] by government entities with a view to improving the efficiency and economy of government operations, as well as to ensure the proper management of these technologies and the systems they serve, their protection from physical harm, and the security and privacy of their information" (Hernon *et al.*, 2002, p. 380).[1]

Rather than identifying and discussing those instruments, this chapter will provide an overview of e-government primarily with reports on observations gathered through monitoring the US government's presence on the World Wide Web (Web) since the late 1990s. That scrutiny has involved the monthly use of link-checking software to track any changes in the addresses of nearly 1000 government home pages and resources. By late 2003, that software had tracked more than 1600 government Web addresses.

The findings from those observations should be factored into further revisions of existing policy instructions, especially those emanating from the Office of Management and Budget (OMB) as it oversees the accomplishment of the E-Government Act of 2002 (P.L. 107-347). Other government entities (e.g., departments, agencies, individual courts, and congressional committees) should reflect on these observations and make some adjustment

---

[1]For a discussion of policy instruments, see Relyea (2003, pp. 369–394) and McClure and Sprehe (2001, pp. 255–292).

in the continued development of their home pages so that future improvements do not compromise progress (two steps forward) with movement in the opposite direction (one step backwards), through the creation of more complex and ever-changing universal resource locators (URLs), more dense Web pages, dead links, the need to insert Flash (Macromedia) graphics on computers to navigate government sites, having to rely on high-speed links to access the content of various government sites, sites that are not readily amendable for use by those with disabilities, and the discovery of pages that are extremely slow to load. In short, a government entity should ensure that those planning documents submitted to Congress under the Government Performance and Results Act (P.L. 103-62) reflect a commitment to achieving the goals set for e-government.

In recent years, government entities have discontinued the printing of numerous publications, relied on the Web as the primary method of information dissemination, treated the Web as more than a mechanism for information dissemination, adapted some features commonly associated with libraries, let their libraries in some instances provide the public with virtual reference service, and have at times treated depository library programs as a complementary method of information dissemination. For example, the Department of Housing and Urban Development (HUD) organizes frequently-requested Web pages according to topics or "bookshelves," while the Patent and Trademark Office encourages users of its home page to be familiar with the collections and services of its depository libraries because patent and trademark searching often requires special expertise. As this chapter illustrates, the Web and e-government have altered the traditional role that libraries play in assisting the public in identifying and retrieving government information. Now, libraries can help their users obtain services and communicate directly with the government, as the public participates in the shaping of public policy. E-government definitely presents both opportunities and challenges to libraries, both depositories and non-depositories.

I. Overview

In the early 1970s, a report of the Commission on the Year 2000 of the American Academy of Arts and Sciences recognized "the conditions contributing to the e-government phenomenon" (Relyea, 2003, p. 379). It suggested that in the new millennium, "despite the growth in the size and complexity of federal programs, the technological improvement of the computer, closed-circuit TV, facsimile transmission, and so on, will make it possible for the federal bureaucracy to carry out its functions more efficiently and effectively than it can today, with no increase in total manpower" (Capron, 1971, p. 307). The report maintained that the use of IT would not be confined to the executive branch. Congress needed "the tools of modern information technology...to create policy and to oversee the Executive." IT would also assist members of Congress in communicating with their constituents and in conducting "up-to-the-minute" polling of public opinion (Brademas, 1971, pp. 319–321).

As policy analyst Relyea (2003, pp. 379–380) notes, the ability of new information technologies to improve government performance and communication did not originate with the dawning of the computer age. Similar predictions were made when the telephone was introduced.

In the 1980s and the early 1990s, national networking—a network of computer networks—emerged. As educator Charles R. McClure and some of his colleagues at Syracuse University wrote,

> while some of the benefits of national networking are difficult to predict, it is clear that the design and implementation of some type of national, coordinated, high-speed network is essential if the United States is to maintain a leadership role in high-performance computing and electronic networking and increase its overall national productivity and competitiveness. (McClure *et al.*, 1991, p. i)

The Clinton administration, through its National Performance Review, advanced the concept of e-government as a way to link the reinvention of government with information and communication technologies (including Internet applications) for the purpose of enhancing access to and delivery of government information and services, improving the internal effectiveness and efficiency of the federal government, and encouraging the entrepreneurial spirit. The administration also supported electronic commerce both within the United States and globally.

Figure 1, which represents a graphic depiction of e-government, shows that it has six parts:

1. Assisting in governance;
2. Supporting emergency response;
3. Engaging in e-commerce;
4. Providing access to information, including records;
5. Delivering services; and
6. Supporting procurement operations.

Each part might extend to one or more of the following audiences: other federal government entities as well as those at a subnational level, the business community, and the public. The public might range from the nation's youth to senior citizens, as well as to librarians, researchers, publishers, and others. For each of these audiences, the intention of government is to be results

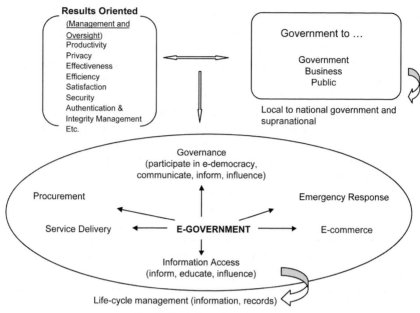

Fig. 1 Depiction of e-government.

oriented, to ensure the security of e-commerce transactions, to offer an efficient channel for providing access to government information, and so on.

Electronic rulemaking is a good example of the role of e-government in assisting in governance. Each year, government entities issue

> thousands of regulations that can affect almost every aspect of citizens' lives—from allowing a fireworks display over the Columbia River...to registering food facilities in light of the potential for bioterrorism. The public can play a role in the rules that affect them through the notice and comment provisions of the Administrative Procedure Act of 1946, as amended. In fact, involvement of the public in rulemaking has been described as possibly "the most complex and important form of political action in the contemporary American political system." However, in order to be involved in rulemaking effectively, the public must be able to (1) know whether proposed rules are open for public comment, (2) prepare and submit comments to relevant decision makers, and (3) access regulatory supporting materials (e.g., agencies' economic analyses) and the comments of others so that their comments can be more informed and useful. (General Accounting Office, 2003a, p. 1)

With this in mind, the General Accounting Office (GAO; now the Government Accountability Office) examined Regulations.gov (http://www.regulations.gov/), which enables individuals to search, view, and comment on proposed regulations issued by any federal entity, and it compared Regulations.gov's coverage to that of selected agencies' home pages. GAO found great variation, with Regulations.gov providing the best—but not complete—coverage of regulations open for public comment. Most often, agency sites did not even mention Regulations.gov "as a commenting option." Another problem was that the location of public comment for regulations could be difficult to locate on a home page. (General Accounting Office, 2003a, pp. 8–18) Although the intent of section 206 of the E-Government Act, which requires agencies, to the extent practicable, to accept public comments on proposed rules "by electronic means," has not been fully met, online rulemaking enables citizens to participate in public policy discussions and the shaping of the resulting decisions.

Returning to Fig. 1, within a country, e-government might span local, state or provincial, and regional governments, as well as the national government. E-government also occurs at the supranational levels, such as through services provided by the European Union. As well, information access must be viewed within the context of the information or record's life cycle, which covers the stages from creation to demise or preservation.[2] Clearly, a diverse set of information policies and policy instruments are results oriented, seek to advance e-government and each part of the figure, and establish a framework for better management of information resources and accountability of IT and e-government.

Privacy and security are issues that cut across all six parts. In the fall, 2003, OMB directed agencies to conduct privacy-impact assessments before developing or changing information systems. Those assessments review how information is collected and used in the organization, and the results of those assessments more than likely will be linked to future funding of a project. In addition, OMB directs agencies to

> develop a plan to make their Web site privacy policies machine-readable—meaning that they automatically provide notification when the site doesn't cover visitors' privacy protection. Agencies must tell Web site visitors when it's voluntary to submit information, how to grant consent for an agency to use voluntary personal data and what their rights are under the Privacy Act. (Michael, 2003, p. 11)

A. Strategy of the Bush Administration

President George W. Bush's fiscal year 2002 management agenda envisions e-government as a way to serve better the public (including persons with disabilities); make government more efficient and effective; reduce government operating costs as well as the expense and difficulty of doing business with the government; and enable the government to become more transparent and accountable. (Willemssen, 2003, p. 4) To achieve these goals, the Bush administration envisioned an expansion of e-government as part of its government-

[2]For a discussion of life cycles, see Hernon (1994).

wide reform effort and as being guided by three principles: the federal government should be (1) citizen-centered, (2) results-oriented, and (3) market-based. (Presidential Memo)[3] Citizen-centered addresses four segments:

1. Individuals: "Building easy to find one-stop shops for citizens—creating single points of easy entry to access high quality government services [and information];"
2. Businesses: "Reduce burden on businesses through use of Internet protocols and by consolidating myriad redundant reporting requirements;"
3. Intergovernmental: "Make it easier for states to meet reporting requirements, while enabling better performance measurement and results, especially for grants;" and
4. Internal efficiency and effectiveness: "Reduce costs for federal government administration by using best practices in areas such as supply chain management and financial management, and knowledge management." (Forman, 2001)

Twenty-five cross-agency initiatives have been selected to achieve both the goals and the guiding principles (General Accounting Office, 2003b).[4] For example, the E-Authentication E-Government Initiative seeks to develop a comprehensive policy applicable across government entities for authentication and identity management. The goal is to eliminate an inconsistent and agency-unique authentication and identity management infrastructure.

The definition of records, as stipulated at 44 *United States Code* 3303, includes agency documents used "in connection with the transaction of public business" and otherwise constitute "evidence of the...functions...operations, or other activities of the Government or because of the information value of data in them." This definition could be applied to all of the material available on government Web sites, a large percentage of which has no print counterpart. The number of digital records that the government produces most likely exceeds the number of records originating in paper form. As well, a number of electronic records were created decades ago but were never sent to the National Archives and Records Administration (NARA) for preservation and public access. These records "may pose challenging preservation problems owing to their age (media deterioration, loss of documentation and other metadata, and obsolesce of data types)" (National Research Council, 2003, p. 2).

---

[3]For additional discussion of e-government as "a critical element in the management framework," see section 2, E-Government Act 2002 (P.L. 107-347, 116 Stat. 2899).

[4]General Accounting Office (2003b) also contains a list of GAO reports related to e-commerce and e-government. See also Implementing (2003).

Another initiative, the Electronic Records Archive (ERA) is a program intended to preserve and make available today's information for future generations; the goal is to ensure that the problem of electronic records management does not remain unresolved.[5] ERA would be enhanced through coordinated efforts with organizations (e.g., OCLC and depository libraries) that "share common interests in digital preservation" (National Research Council, 2003, p. 3). One such partnership involves the Government Printing Office (GPO), which would have "responsibility for public access and preservation of the records available on…GPO Access" (Reed, 2003a, p. 47). In essence, GPO becomes an "affiliated archive" (Reed, 2003a, p. 47) as it retains electronic records, such as copies of the *Federal Register* and the *Code of Federal Regulations*; however, NARA still retains legal custody.

B. The Impact of Terrorism on E-Government

Since the unfolding of the US government on the Web under the Clinton administration, greater amounts of government information have become publicly available. Anyone with a computer or access to one can browse vast storehouses of information and select the few items most relevant to his or her needs, without having to disclose personal information. By using publicly available terminals, it is possible to hide one's identity at least for a while. However, following the terrorist attacks of 9/11, the government reevaluated a number of its Web sites and, in some instances, removed content and reorganized the site.[6] The question is, "At what point does the removal of

---

[5]Complicating the realization of the goal of e-records management, "NARA still advises agencies to print their e-records." See Sprehe (2003), p 40. For an overview of digital preservation, see the National Digital Information Infrastructure and Preservation Program, http://www.digitalpreservation.gov/

[6]OMB Watch, a non-profit, public interest organization, has a page (lasted updated in May 2002) that identifies "information restriction policies" and "information removed from agency websites" (see http://www.ombwatch.org/article/articlereview/213/1/04/). Complementary information is available on the home page of the Federation of American Scientists, a nonprofit, tax-exempt organization. One of the programs available is the Project on Government Secrecy, http://www.fas.org/sgp/. See also French, (2003), p. 9. In the fall 2003, NARA announced that it was "reevaluating access to some previously open archival materials, and reinforcing established practices in screening materials not yet open for research." It cited as authority for these actions the exemptions listed in the Freedom of Information Act, Department of Justice instructions (exemption (b) (2)), deeds of gift, the Presidential Recording and Materials Preservation Act, the Presidential Records Act, and various executive orders. See http://www.archives.gov/research_room/whats_new/notices/access_and_terrorism.html.

In February 2004, OMB Watch noted that the U.S. Office of Special Counsel (http://www.osc.gov/), an independent agency charged with protecting the workplace rights of federal employees, removed reference to sexual orientation from its Web site and other agency materials. The removal represents a reversal of agency policy since 1975. See "Office of Special Counsel Scrubs Website" (2004).

information from the public domain—what information and for what purposes—inhibit democracy and the accountability of government to the public?" As policy analyst L. E. Halchin notes, "the removal, or withholding, of what was once considered public information from agency Web sites may thwart the promise of e-government...[T]he ongoing debate over the removal of information might detract from the luster of e-government" (Halchin, 2002, p. 249).

When removal extends to scientific information unrelated to security matters, it should be remembered that "science is a collective endeavor... [and] Science increasingly...[is] an international endeavor" (Vest, 2003, p. 23). "Restraining scientific publication and the international exchange of information could adversely affect public health by inhibiting scientific research and medical progress" (Atlas, 2003, p. 15). Thus, removal of information from the Web, scientific and other, involves a delicate balance between providing a means to retard terrorist activities and the continuing advancement of research, progress, and knowledge.

Another complication is that different executive branch entities have removed their coverage of the Freedom of Information Act (FOIA) from a prominent place on the opening site of their Web site. In some instances, the information seeker must check the site map or conduct a search of the site to locate coverage of the Act and any declassified records that the agency displays for public consumption. The entities have reviewed the types of declassified records they release through their home page.

Unrelated to the terrorist attacks of 9/11, in December 2001, a federal district court judge ordered the Department of Interior to shut down its Web sites, including that of the Bureau of Indian Affairs (BIA). The shut down was related to the department's mismanagement of funds intended for American Indians. The department's home page was reopened in 2002, but that of the BIA remains closed to date (September 2004). Consequently, contact with the BIA and its resources depends on channels other than the Web.

C. The Evolution of Web-based Government Information

Section 205 of the E-Government Act of 2002 instructs federal courts to provide access to certain types of information, including contact information, local rules, standing or general orders, docket information, written opinions, documents filed in electronic format, and other information that a court deems "useful to the public." The Web environment also provides access to more government information resources (e.g., publications, images, records, and datasets) than were available in a paper environment. At the same time, federal government entities deliver online services (e.g., agency library

The US Government and E-Government: Two Steps Forward, One Step Backwards? 129

collections and reference services online, subscription and e-mail notification services, online retail, online forms and instructions, and enabling people to arrange for the receipt of benefits) and facilitate the procurement of online goods and services, as well as the efficient exchange of information, goods, and services with subnational governments. Online retail, for instance, generates millions of dollars annually.

Those responsible for improving and maintaining publicly-accessible federal Web sites are presenting more content by means of Web applications and services first deployed by the private business sector, such as Macromedia-based Flash animations and Java language-based modules. JavaScript-based rollovers are commonly deployed on federal Web sites, which may provide a site search engine to facilitate navigation. The intent is to enrich the public's visit to the Web site with attractive presentations and easily accessible content. Dynamic HTML is common; deployment of XML is increasing, which is intended to improve the sharing and delivery of content among government entities, with commercial and industrial contractors and organizations, and with consumers. For example,

> Beginning in 1997, the House and the Senate, along with the other Legislative Branch agencies, began an investigation of the use of SGML and later XML as a data standard for the exchange of legislative documents. By December 2000, the Committee on House Administration and Senate's Rules Committee adopted XML as the primary standard for the exchange of legislative documents between the House, Senate, and other legislative branch agencies. The Legislative Branch including the House, Senate, Government Printing Office, Library of Congress, Congressional Budget Office, and the General Accounting Office maintain coordination in terms of the Common Tag Library for legislative documents (Congress).

Therefore, the government, through agencies such as the National Institute for Standards and Technology (NIST) is not just a convener and steward of electronic standards and guidelines; it is also a consumer.

1. Portals

Realizing that access to Web-based government information is comparable to finding one's way through a maze of undeterminable size and shape, the Clinton administration sought to create a single search engine designed to integrate nearly all federal government home pages. That search engine, WebGov, evolved into the portal FirstGov; a portal is a multifunctional Web site that usually includes Web directories, indexes, constituent services, and links to other appropriate Web-based resources. In essence, a portal guides users through that maze by creating sites that, it is hoped, provide one-stop shopping (Hernon *et al.*, 1999, pp. 211–228). Because "at least 70 percent of FirstGov visitors are citizens, and most of these visitors are looking for help with services such as applying for social security or changing an address," the portal has made the citizen tab into its default home page (Frank, 2003, p. 40). This change to the portal reflects the administration's *three clicks to service or information strategy*, which stipulates that users of FirstGov should only have to follow three links to find the information or service they seek.

Reporter Ed McKenna notes that, by "hosting various enterprise applications for both public and internal use, providing tools for online collaboration, and serving as user-friendly front ends to vast stores of distributed information, portals are becoming mission critical for many agencies" (McKenna, 2003, p. 32). These portals might convey service initiatives, provide information resources, and further the accomplishment of e-governance. Examples of such portals are

- Recreation (http://www.recreation.gov/), which offers a single point of access to information about parks and government recreation areas;
- Gov On-line Learning Center (http://www.golearn.gov/), which provides a single source for online training of federal employees;
- Recruitment One-Shop (http://www.usajobs.opm.gov/), which assists applicants in finding employment in the federal government;
- Geospatial One-Stop (also known as the geodata.gov portal; http://www.geodata.gov/), which stores data collected by federal, state, and local governments so that users of geographic information systems (GIS) can readily find data and then combine, enhance, and analyze those data;
- Grants.gov (http://www.grants.gov/), which provides cross-departmental and agency access to federal grants;
- FEDSTATS (http://www.fedstats.gov/), which provides access to statistical data from more than 100 federal agencies; and
- GPO Access (http://www.gpoaccess.gov/), which "has an average of 32 million documents [that are] downloaded each month, and that number is growing" (Frank, 2003, p. 41).

In addition, a number of agencies, as well as courts, have developed electronic dockets, which "are formal inventories of materials making up the record in a proceeding…[and] as a practical matter the docket defines the record" (Perritt, 1995). Such dockets encourage greater dialogue or communication directly among stakeholders, citizens and other user groups, and agencies at national and subnational levels. Both e-governance and e-services might have an outreach and education component. Government entities might maintain electronic mailing lists to provide announcements as well as access to new publications and policy changes.

2. Redesigning Home Pages

Many government home pages contain a link to FirstGov and have been redesigned to resemble that portal and to make it easier for the public to navigate the wealth and diversity of available information. For example, the Small Business Administration, which launched its home page in 1992, has served more than 1.2 million visitors to its site each week; a site that offers more than 50,000 publications! The redesign involved removing "excessive jargon and confusing terminology" while "adding tutorials and training to help users learn how to do business with the federal government," and creating specific "information categories designed to guide users through the small-business process: starting a business, financing a business, managing and growing a business, business opportunities, and disaster assistance" (Reed, 2003b, p. 36).

In an attempt to simplify access to its Web resources, the National Aeronautics and Space Administration (NASA), which "has more than 3000 Web sites hosting 4 million pages of information," has begun consolidating content from a number of those sites into its main site, http://www.nasa.gov/home/index.html (Hardy, 2003, p. 30).[7] Consequently, users will not have to navigate so many sites or know which specialized sites contain the information they want.

As a result of such efforts, some government Web sites "score high on user satisfaction survey[s]." The National Women's Health Information Center of the Department of Health and Human Services (http://www.4women.gov/) scored the highest among government sites on one satisfaction survey. In fact, that site "scored higher than several prominent private sites and on a par with Amazon.com" (Daukantas, 2003b, p. 68).

D. Blurring the Role Between the Public and Private Sectors: Government Expands Web Dissemination

E-government is forging partnerships and alliances with the private sector and government agencies (even those at subnational levels of government). As a result, more, better organized, and better displayed government information and services are readily available. Furthermore, a number of entities tailor access on their home pages to specialized audiences, such as teachers, businesses, publishers, and youth.

Although the information and records provided are mostly current, they might also be historical. For example, the predecessors to the *Congressional Record* are available digitally up to 1873 and the *Congressional*

---

[7]See also Lisagor (2003, pp. 36–37).

Record is available on government portals since the early 1990s. That gap from 1873 to the early 1990s is one for the private sector to close, if it so chooses. The State Department series, the *Foreign Relations of the United States*, provides declassified foreign policy records back to 1861; more recent volumes in this series are also available digitally through the department's home page (http://www.state.gov/). Agency Web sites might also contain specialized software to make some machine-readable information produced decades ago available to whoever wants it. For example, the US Geological Survey (USGS) offers GEODE (http://dss1.er.usgs.gov/) and the Environmental Protection Agency's (EPA) Office of Science and Technology provides BASINS (Better Assessment Science Integrating Point and Nonpoint Sources, http://www.epa.gov/epahome/gis.htm).

Government entities are cognizant that their Web visitors use a variety of workstation platforms (Intel and Apple), browsers (Netscape and Microsoft Internet Explorer), and Internet access speeds (telephone, cable, and digital subscriber line (DSL)), as well as modems and local area networks, and workstation-installed software productivity applications (e.g., Microsoft and Corel office suites). As a result, government Web sites strive to meet individual user needs by providing users with alternatives and choices for viewing information and downloading files based on the speed of their Internet connection and installed viewer. An example is the "Space Research" page of NASA's Office of Biological and Physical Research (http://spaceresearch.nasa.gov/fun_learning/robot.html), which provides visitors with the option of downloading video clips via dial-up or broadband. Dial-up video clip files are usually smaller and have less resolution than the larger, higher resolution broadband files.

Portals cannot provide access to all information, records, and services that the government offers or plans to offer. Furthermore, there is great variation among government entities about which information resources and services they provide. E-government users must often explore different sites in the pursuit of relevant information, records, and services. As they navigate government on the Web, they will find examples such as the following:

- A fully-functional advanced search with options (search by article or book title, the search term in an abstract, keywords, authors, etc.) (Department of Transportation, Bureau of Transportation Statistics, *TRIS Online*, http://tris.bts.gov/sundev/search.cfm).
- An opportunity for users to establish a customized version of Export.gov—the US Government Export Portal—so that they may receive information concerning exports, international markets, and international trade (need to

set up a password, Export Gov Community Registration, http://ita-webhost1.ita.doc.gov/soap2/register.jsp).
- An online guide for installing GEODE that assists users through a difficult procedure (USGS, http://dss1.er.usgs.gov/help).
- An excellent explanation of what a.pdf file is (Library of Congress, http://thomas.loc.gov/tfaq15/pdfhelp.html).
- Access to "a searchable library of transportation specifications from across the country. It includes emerging specifications in the areas of quality assurance, performance-related, warranty specifications, and other innovative specifications. The site features a discussion forum to enhance communication and feedback among the community of users" (Federal Highway Administration, http://fhwapap04.fhwa.dot.gov/index.jsp).
- Access to PURLs (persistent uniform resource locator) for free and convenient access to full-text and bibliographic records of Department of Energy research and development reports in physics, chemistry, materials, biology, environmental sciences, energy technologies, engineering, computer and information science, renewal energy, and other subjects (Office of Scientific and Technical Information, http://www.osti.gov/bridge/).

A noteworthy development occurred in October, 2003, when the National Institutes of Health accepted 14 grant applications electronically. By October 2004, it expects to handle its R-01 grants in a similar manner.

Other examples of what government entities are doing on their home pages include *webcasting*, or audio and video sent through the Web. A popular type of webcast is *streaming*. When an audio and/or video file is *streamed*, it means that the user can hear or see the file without having to wait for the entire file to download. Congressional committees often engage in webcasting as does HUD when it provides live coverage of training and public events through its home page.

Government Web sites also provide users with more interactive functionality, enabling them to create, modify, or customize available government information to meet their specific and individual needs. For example, the *National Atlas Online* (USGS, http://www-atlas.usgs.gov/atlasvue.html), which uses Shockwave, requires that frames be enabled so that users can customize maps interactively within a user's Web browser. *Dumptown Game* (EPA, http://www.epa.gov/recyclerity/gameintro.htm), which employs Macromedia's Shockwave, enables users to watch the image move and change as they interact with the program as the hypothetical city manager of Recycle City. The EPA also has *EnviroMapper* (http://maps.epa.gov/enviromapper/), which provides users with interactive GIS functionality using EPA spatial data for the conterminous United States.

II. Issues

This section highlights five issues: (1) restructuring and consolidating a major educational program; (2) Web privacy; (3) the extent of use, misinformation, and disinformation; (4) data quality; and (5) section 508 compliance. While these issues tend to represent progress, or steps forward, some readers might see certain aspects as impeding the furtherance of

- public participation in e-government and the availability of information (providing accountability, informing the public, and enabling people to lead better and more productive lives); and
- the introduction of new services (serving the public better and in new and creative ways).

A. ERIC Restructuring

Despite the innovations highlighted in the previous section, there is some concern that not all of the services that the government provides online actually advance e-government; in fact, they might represent steps backwards. A good example occurred in spring 2003 when the Department of Education announced a massive restructuring of the Educational Resources Information Center (ERIC) by eliminating the clearinghouse and many of its user services. The department also announced its intent to change the content of, and the number of journals covered by, ERIC's database. Since the announcement these clearinghouses have been consolidated under one contractor and that contractor manages the electronic publishing, dissemination, and archives collection. The contractor is also designing a Web site that will "make information accessible in a user-friendly, timely, and efficient manner" (Educational Resources Information Center, 2003). "Many researchers conceded that the current system has redundancies and can be difficult to navigate electronically. But some worry that the proposed streamlining would involve elimination of valuable services, materials, and expertise." Furthermore, some of the material deleted from coverage in ERIC may not be readily accessible elsewhere, thereby "curtailing access to information" ("Government Proposal May Curtail Access to Data," 2003, p. 8). As is evident, educators and others will monitor the new ERIC to determine if it represents a step forward or backwards.

B. Privacy

Government entities might gather and store data on individuals who use their home pages; however, any data collected should not impinge on the public's right to privacy as recognized in the Bill of Rights and existing statutes and regulations. Any analysis that government entities do with the data they collect should be at the aggregate, not individual, level. Furthermore, any data that these entities collect should not monitor individuals' repeated use of a Web site or Web page. When those entities use cookies—small computer files placed in a Web site visitor's hard disk that track that person's travels on the Web to determine who visited the site recently and how that person got there—those files should not gather invasive information about people and their online use, nor should they track search behavior without user consent.

OMB lets government entities use *session cookies* that expire once the user closes the Web browser at the end of an online session, but prohibits them from employing *persistent cookies* that only expire after a specific time. Thus, it is important for government entities to explain their policy about any use of cookies and the type used on the opening screen of their home page. Many do not do this, however. Thus, does the use of cookies represents a step forward or backwards?

In Memorandum M-00-13 issued on June 22, 2000, OMB reminded each agency of its requirement "to establish clear privacy polices for its web activities and to comply with those policies" (Office of Management and Budget, 2000). Furthermore,

> Particular privacy concerns may be raised when uses of web technology can track the activities of users over time and across different web sites. These concerns are especially great where individuals who have come to government web sites do not have clear and conspicuous notice of any such tracking activities. "Cookies"—small bits of software that are placed on a web user's hard drive—are a principal example of current web technology that can be used in this way. The guidance issued on June 2, 1999, provided that agencies could only use "cookies" or other automatic means of collecting information if they gave clear notice of those activities.
>
> Because of the unique laws and traditions about government access to citizens' personal information, the presumption should be that "cookies" would not be used at Federal web sites. Under this new Federal policy, "cookies" should not be used at Federal web sites, or by contractors when operating web sites on behalf of agencies, unless, in addition to clear and conspicuous notice, the following conditions are met: a compelling need to gather the data on the site; appropriate and publicly disclosed privacy safeguards for handling of information derived from "cookies"; and personal approval by the head of the agency. In addition, it is federal policy that all Federal web sites and contractors when operating on behalf of agencies shall comply with the standards set forth in the Children's Online Privacy Protection Act of 1998 with respect to the collection of personal information online at web sites directed to children. (Office of Management and Budget, 2000)

Agencies have complied by making an effort to inform their visitors. For example, NASA's policy states that

NASA uses advanced technologies as part of its core mission to discover and inform. Cookie technology may be implemented at some NASA Web sites. At no time is private information you have given us, whether stored in cookies (persistent) or elsewhere, shared with third parties that have no right to that information. If you do not wish to have persistent cookies stored on your machine, you can turn them off in your browser. However, this may impact the functioning of some NASA sites.

We may collect and store information for statistical purposes. For example, we may count the number of visitors to the different pages of our Web site to help make them more useful to visitors. This information does not identify you personally. We automatically collect and store only the following information about your visit:

1. The Internet domain (for example, "xcompany.com" if you use a private Internet access account, or "yourschool.edu" if you connect from a university's domain) and IP address (an IP address is a number that is automatically assigned to your computer whenever you are surfing the Web) from which you access our Web site;
2. The type of browser and operating system used to access our site;
3. The date and time you access our site;
4. The pages you visit; and
5. If you visited this NASA Web site from a link on another Web site, the address of that Web site.

The information that you provide on a NASA Web site will be used only for its intended purpose, except as required by law or if pertinent to judicial or governmental investigations or proceedings. (National Aeronautics and Space Administration)

The US Mint's home page includes a link to its cookies policy by using an image of a chocolate chip cookie. However, in their posted privacy policies, the Web sites of most government entities (including NASA and the US Mint) clearly state that it is the responsibility of the visitor to either turn off the ability to accept cookies in their browsers, or, as in the case of the US Mint, to "delete any US Mint.gov cookies from your hard drive" after leaving the site (see US Mint). Nonetheless, they fail to offer information about how to turn off the application or how to delete cookies from one's hard drive. This issue comes important if government entities, contrary to OMB's policy, use persistent cookies.

C. Extent of Use, Misinformation, and Disinformation

In its report, *The Rise of the E-Citizen: How People Use Government Agencies' Web Sites*, the Pew Internet & American Life Project estimated, for instance, that

- "68 million American adults have used government agency Web sites... They exploit their new access to government in wide-ranging ways, finding information to further their civic, professional, and personal lives. Some also use government Web sites to apply for benefits, engage public officials,

and complete transactions such as filing taxes.
- 42 million Americans have used government Web sites to research public policy issues.
- 23 million Americans have used the Internet to send comments to public officials about policy choices.
- 14 million have used government Web sites to gather information to help them decide how to cast their votes.
- 13 million have participated in online lobbying campaigns.
- Most government Web site visitors are happy with what they find on the sites; 80% of them say they find what they are seeking on the Web sites." (Pew Internet & American Life Project, 2002, p. 2)

Not surprisingly, when asked about any future terrorist attacks on the nation's homeland, those responding to different poll indicated that in such an eventuality they would rely on television and radio, not government Web sites, for up-to-date news coverage. They would also expect government to provide the media with reliable information for inclusion in its reporting (Hasson and Holmes, 2003).

With so many people using US government on the Web, it would seem that they place trust in government and the resources provided. Furthermore, with portals such as FirstGov providing access to resources across branch and level of government, there is an effort to create transparency of government. Finally, there is a belief that "government Web users, more than other Internet surfers, tend to be affluent and educated" (Pew Internet & American Life Project, 2002, p. 2). To change these demographics, the number of government home pages containing resources in languages other than English has increased over the past couple of years. As well, the three branches of government have more than 70 sites aimed at primary and secondary students, parents, and teachers (Hernon et al., 2003, pp. 353–375). Yet, some members of the public now question the reliability of information presented on some executive branch home pages and they charge that such information reflects the conservative ideology of the Bush administration (Hernon et al., 2003, p. 18). If the administration is not careful, there could be an erosion of public trust in e-government.

Finally, misinformation applies to honest mistakes and information that computer hackers post on government home pages, whereas disinformation relates to the intent of government to deceive others, often governments hostile to the United States and terrorist groups. Much government and other information presented on the Web is unfiltered, and there may be a desire to deceive or confuse—to shape and sway public opinion in the United States and elsewhere. The Web is a means to convey information, data, and messages—truthful, deceptive, or somewhere in between—to an audience.

D. Data Quality

The Treasury and General Government Appropriations Act (P.L. 106-554) directed OMB to issue guidelines that ensure and maximize "the quality, objectivity, utility, and integrity of information (including statistical information) disseminated by Federal agencies in fulfillment [of]...the Paperwork Reduction Act" (section 515). In 2001, OMB issued the guidelines, which were then revised in September 2003. In response to criticisms raised during the public comment period, OMB stated that "it does not envision administrative mechanisms (appeals about the quality of specific datasets) that would burden agencies with frivolous claims. Instead, the correction process should serve to address the genuine and valid needs of the agency and its constituents without disrupting agency processes" (see Office of Management and Budget, http://www.whitehouse.gov/omb/ferdreg/reproducible.html). Undoubtedly, nobody would object to increased efforts to ensure that the government only disseminates data of the high quality. However, charges of inferior quality should neither inhibit public access to government information nor interfere with existing rulemaking processes. Despite OMB's assurance, there is concern that the guidelines might be misused to delay, manipulate, and influence the outcome of agency reviews.

E. Section 508 Compliance

Amendments to the 1973 Rehabilitation Act, enacted in 1986, created Section 508, which became operational in June 2001, and "requires federal departments and agencies to ensure that their development, procurement and maintenance of electronic and information technology allows people with disabilities—both employees and the public—to have access to information and data comparable to those without disabilities"[8] (Reed, 2003a, p. 21). At first, agencies did not understand their responsibilities under the new law. Many of them still "do not know how to comply with...[it]" (Reed, 2003c, p. 21). Yet, if people with disabilities cannot benefit from all six parts of e-government (as identified in Fig. 1), the resulting barrier represents one step backwards. Unless any redesign of government Web sites complies with Section 508, those with disabilities will not have access to the diverse content of government on the Web, thereby increasing the digital divide.

---

[8]See also Daukantas (2003a, p. 38), for tips on "improving site accessibility without adding layers of technology."

III. Definite Barriers to Information Access (One Step Backwards)

In preparing and updating the content of *US Government on the Web* (Libraries Unlimited, 1999; now in its third edition), we have identified a number of features on government Web sites that would further public access. However, there are significant inconsistencies among Web sites as to the presence of these features (e.g., site maps and search engines that permit advanced searching) (Hernon *et al.*, 2003, p. 24). Barriers—be they physical, economic, or technological—impede e-governance and the flow of information and services to citizen, businesses, and national and subnational government. "These barriers may be actively imposed by government, or they may be allowed to continue simply through lack of action by government" (Cullen and Houghton, 2000, p. 244). Furthermore, these barriers hinder progress and, in some instances, are counterproductive—they clearly represent steps backwards.

By using link-checking software, we have monitored the extent to which Web addresses listed in *US Government on the Web* are unstable (see Table I). Dead links are URLs that no longer function, whereas, with redirected links, the URL has changed. However, the user is redirected from the old URL to the new one. Most redirected URLs are temporary, and later become dead links. Additionally, redirected URLs do not update browser bookmarks.

The numbers and percentages would be much more dramatic if we had included the number of changes made to URLs at the time of the page proof stage of production for each edition. Clearly, for whatever reason, government entities frequently revamp their Web sites and pages, and the presentation of their digital information resources. The problem is that, over time (better measured in years than months), URLs change as government entities expand their Web-based content. This results in a revision of URLs on the Web pages within the site, changing content as government entities revise their mission (e.g., those entities impacted by

Table I
Dead and Redirected Links in *US Government on the Web* (Libraries Unlimited)

	Total links	Dead links	Redirected links	Percentage
From 1st edition (1999)	920	253	279	57.8
From 2nd edition (2001)	1272	99	234	26.2
From 3rd edition (2003)[a]	1668	16	80	5.8

[a] These numbers are current as of December 3, 2003.

homeland security) and as Web sites evolve in applications deployed (e.g., improved graphics and changes in standards applied, such as XHTML replacing HTML). Thus, as the content of the second and third editions ages, the percentages will become more dramatic and perhaps equal those of the first edition.

Table I suggests that there might be a need for government (in particular OMB) to develop performance measures to determine an acceptable percentage of dead and non-functioning redirected links. In some sectors, an error rate of 1% is acceptable. The percentages listed in Table I far exceed this. OMB, as charged by the E-Government Act of 2002, should investigate this issue as it considers the impact of changed URLs, dead links, and non-functioning redirected links on long-term public access to the content of Web sites. It would seem that dead, and non-functioning redirected, links pose the greatest barrier to public access.

As federal government Web sites evolve to include more information or attention-attracting features such as Flash (Macromedia) graphics, the complexity of Web addresses (URLs) increases. For example, many federal Web pages now end in extensions such as ".asp" and ".jsp" rather than the older and more common ".htm" or ".html" extensions. "JSP" extensions refer to Java Server Pages technology, while "ASP" refers to Active Server Pages. These modules are intended to extend the capabilities of a Web server to provide dynamic Web scripting/programming that offers platform independence, enhanced performance, ease of administration, and, most importantly, ease of use. However, to take advantage of these applications, the deployed Web address should include the extension in order to inform transparently the user's browser of the need for specific plug-ins to execute the module.

Additionally, as federal Web sites increase in the number of pages available and services offered, the URLs are becoming physically longer and specific in an effort to provide easier navigation for the user to the specific information wanted. While the intent is sound, the resulting URLs are becoming increasingly long and undecipherable. Furthermore, the URLs are often revised as government Web managers continue to reorganize their sites to improve site navigation which is increasingly important as the Web sites expand both in the content and presentation, and to improve site management, including its reliability for user availability.

IV. A Modest Research Agenda

The research involved in the collection of data relevant to the analysis of information policies, and the improvement and the delivery of services and information, related to e-government has relied on multi-method data collection. These methods include the use of survey, in-person and focus group interviews, content analysis, transaction log analysis, usability studies, obtrusive evaluation, eye tracking studies, and so forth. Additional research might:

- Expand the tool chest of methods (e.g., use verbal protocols such as think aloud/think after protocol).
- Investigate how individuals with disabilities navigate, select, and use government Web sites and their content.
- Apply a revised SERVQUAL instrument from marketing to determine citizen expectations of government services and information dissemination. SERVQUAL deals with the gap between citizen expectations and the actual delivery of services and information.
- Conduct more detailed examinations of users of government home pages, their use patterns, preferences, and satisfaction. For example, who uses the home pages of sites aimed at the nation's (or global) youth? To what extent are resources in non-English languages used and by whom?
- Determine the extent of errors (e.g., broken links) on government Web sites and compare the results to a study in the United Kingdom that found UK sites "have, on average, 600 errors each" ("Report Bashes U.K. Government Web Sites," 2003, p. 13).
- Investigate the principle of three-click access proposed by the Bush administration. The Bureau of Economic Analysis claims, "everything [on its home page] is reachable with two clicks of the home page" (Daukantas, 2003c).[9] This claim, as well as that for FirstGov that desired information or a service should be reachable within three clicks, should be tested. Such claims could be converted into performance or other measures that reflect a citizen perspective.
- Determine how many people currently use the GPO depository library program and for what purposes. How does the public use GPO Access to locate and retrieve information? When people seek access to government information remotely or off-site, do they use the depository home page? If yes, for what purposes? How satisfied are users with depository library services and electronic links? How do depository libraries help to advance e-government as depicted in Fig. 1?

---

[9]On its home page, the Bureau of Economic Analysis answers "questions about using this web site." This page is most useful for anyone wanting to know about the purpose and usability of the site (see http://www.bea.doc.gov/bea/faq/web/FAQ.htm).

Government entities within all three branches of government place large quantities of statistical data on their Web sites as electronic tables in column format. Educators Gary Marchionini and Xiangming Mu examined how people use "highly compressed and highly structured" e-tables, and they designed and tested a Web-based browser to assist the public in using these tables. Figures 2–7 of their article plot eye movement for tasking a simple lookup, a comparison, and trend analysis (Marchionini and Mu, 2003). Eye movement studies, as well as other types of data collection, could be applied more broadly to electronic tables and to have people navigate government portals. If they encounter page after page of screen listings presumably relevant to their search, how do they decide which items to select? Do they use only the first screen (e.g., of FirstGov), or do they know how to read all of the entries (even if 500–1000 items are listed) and how to separate perishable (e.g., press releases) from other kinds of information resources (e.g., reports)? Also, what prototype interface tools can be developed to simplify information identification, retrieval, and use?

Accenture, a global management consulting and technology services company, has conducted a number of studies on e-government in the United States and elsewhere. Those studies provide comparative evidence of the emergence of e-government globally and suggest that e-government initiatives develop in five distinct stages: online presence, basic capability, service availability, mature delivery, and service transformation.[10] With more government entities apparently engaged in service transformation, are they additional stages? If Accenture's characterization is correct, are there differences in the mature delivery and service transformation stages within the Web sites of a government entity, across entities, and across branches of government? If there are differences, what is their significance?

The Benton Foundation released a report, *Achieving E-Government for All*, which documents that

> information on most government websites is skewed to the needs and abilities of highly educated English speakers. For low-literate populations, the Web remains an untapped resource. People with disabilities, such as those with visual impairments, continue to struggle with government websites that don't address their needs. (Benton Foundation, 2003)

---

[10]See eGovernment Leadership: Engaging the Customer and other research reports produced by Accenture (http://www.accenture.com/xd/xd.asp?it=enweb&xd=industries/gove.).

Furthermore, "inaccessible, unreadable government websites affect real people—those who often can no longer find what they need in the offline world, as governments migrate critical information and services to cyberspace." The report also notes that "half of American are reading at the eight-grade level or lower," whereas "many Web sites require an eleventh grade reading level." Regarding accessibility of Web sites, "47 percent of federal sites satisfied the W3C [World Wide Web Consortium] standard of accessibility [for priority level one]" and "22 percent...were in compliance [with Section 508 guidelines]" (Benton Foundation, 2003, pp. 2 and 3).

These statistics suggest that researchers might apply tools, such as the online Bobby service (http://bobby.watchfire.com) to test different Web sites within executive and legislative branches. For example, the White House home page (http://www.whitehouse.gov/) "does not yet meet the requirements for Bobby AAA Approved status." When home pages identify target audiences (e.g., the general public and kids), and when pages provide information in languages other than English, what is their rating and how readable are they? How can the information compiled be used to improve the rating of these sites?

V. Implications of E-Government to Libraries

The increasing emphasis of the federal government on e-government initiatives and efforts results in a shift from being a limited distributor of information products and services to being a 24/7/365 direct information provider. In the past, the government has used (but not exclusively) the GPO for printing services and depository library programs (e.g., those of the GPO, Patent and Trademark Office, and the Bureau of the Census) to provide the public with physical access to its information products. However, e-government programs have decentralized the accessibility of government information from fewer than 1500 GPO depository libraries and one physical government bookstore to the millions of consumers with access to a computer and the World Wide Web. E-government enables government entities to be citizen-centered when it comes to information distribution and dissemination. In fact, FirstGov was designed as a portal to e-government enabling users to interact with a government information provider directly through its Web site.

Such ubiquitous decentralization is not without its problems. Since almost any federal entity can literally publish almost anything it compiles, there may be a reduction in quality control concerning content and presentation. Permanent accessibility to available information is questioned—who is responsible for preserving the content if it is in electronic format and not distributed to an appropriate source, such as a library, for physical accessibility and archiving? Web pages and their content disappear without warning, and Web-based addresses for documents and services are often revised without proper re-direction. Web sites reflect the institutional and organizational culture of their maintainers—navigation may become unnecessarily complicated as the site's content and services expand while the products and services are inadequately indexed so as to be easily lost while using internal site search engines. As a result, users are left with an increasing maze of navigation and content that renders their information seeking frustrating and futile.

Web content does not necessarily adhere to the traditional model of the life cycle of government information. New information may never be posted to an agency's Web site; a document may be deemed to be "internal" and not for public consumption, unavailable because of national security, or not fitting with the politically driven image of the information producer/provider. Flawed information may be quickly removed and not replaced. Information may only be available for a short time on the Web, and its print counterpart never produced. Information previously difficult to destroy because of its distribution to a multiplicity of physical facilities may be irretrievably lost with a few keystrokes.

Outsourcing of federal information becomes easier. Third parties seek to protect their investments in adding value to federally produced information. For example, in November 2002, the Department of Energy's Office of Scientific and Technical Information discontinued PubScience, an indexing and abstracting service, because private-sector companies such as Scirus (http://www.scirus.com/) and Infotrieve (http://www4.infotrieve.com/index.asp) offered comparable, and competitive, services.

Libraries have always added value to federal information by acquiring, cataloging, shelving, and otherwise preparing and maintaining federal information for user accessibility. Value-added library services are necessarily shifting from locator, shelver, and preserver to "access facilitator" as federal information continually migrates from ink on paper to electronic formats. Permanent preservation of information is certainly a long-term availability issue that is important to future research needs, the individual user, businesses, and government itself. However, the management surrounding the federal government's shift from a traditional information cycle to the electronic cycle is a larger cultural, research and accessibility issue than libraries alone can address.

VI. Conclusion

A. Information Policy

Given the complexity and the sheer size of the federal government, one purpose of e-government is to be citizen-centered through the creation of greater transparency or structures that allow the public, government, and businesses to track issues, services, and information throughout the entire organization and across organizations. As Robert D. Carlitz, Executive Director of Information Renaissance (http://www.info-ren.org/), and Rosemary W. Gunn, National Project Manager of Information Renaissance, explain, transparency is "more than an E-government buzzword" or a "good government goal"; for instance,

> Regulated entities find it easier to do business when the process of regulation is more predictable. Agencies themselves have a need to organize and access information across internal agency boundaries. When information is not readily available, an agency is apt to be less efficient in assessing and reacting to its environment, including its ability to defend or enforce existing regulations, or to incorporate stakeholder viewpoints in new rules. (Carlitz and Gunn, 2002, p. 392)

However, despite the improvements in government Web sites and the intention to make government departments and agencies more accountable for their results (see Fig. 1), e-government is not entirely a continuous or unabated progression toward the goal of improved information access, services, democracy and governance, and e-commerce. The numerous changes in, and the length and complex of, URLs, complicate the location and retrieval of needed information. There may be dead links and typographical errors on Web sites that call into question the accuracy and trustworthiness of the information provided. Other steps backwards include the fact that efforts to simplify access to the information and records on a home page may be counterproductive. Given the Bush strategy and its application by some agencies, we might ask, "How much material can or should be retrieved within three clicks of the mouse?"

The E-Government Act of 2002 established an Office of Electronic Government (OEG) within OMB and charged it to work with the Office of Information and Regulatory Affairs and other offices within OMB. OEG has a role in ensuring "access to, dissemination of, and preservation of Government information" (Section 3602(e)(5)) and in providing "overall leadership and direction to the executive branch on electronic Government" (Section 3602(f)(3)). Any steps backward should be labeled as one of the "disparities in access to the Internet" (Section 215) and corrected.

Finally, the war on terrorism influences the extent to which all aspects of Fig. 1 can be fully achieved. Assuming the availability of sufficient funds, the full vision of e-government cannot be achieved as long as there is no attempt to balance (or to discuss what the proper degree of balance is between) open and closed access. To what extent does scientific, economic, and technological progress, as well as an informed citizenry, necessitate an even-handed balancing of the scale? Does the war on terrorism serve as an excuse to expand the amount of information and records outside public scrutiny? Clearly, policy makers, together with concerned public interest groups, should enter into a discussion of Fig. 1 and the proper balance between open and closed access to government information and records.

B. Role of Library Community

Despite the efforts of the government since the 1900s to make e-government more transparent, access to government information and services, and the range of topics covered in Fig. 1, can be very difficult. People searching for government information resources need a good understanding of how the government works, the structure of government, terminology (e.g., the difference between a *report* and *committee print*, a *record* and *information*, a *statute* and a *regulation*, and the *Statutes at Large* and the *United States Code*), the role that different agencies play (e.g., the GAO as the investigative agency for Congress), and the realization that government Web sites might end with an extension other than .gov or .mil.

Librarians, more than those serving in a depository collection, can play an important role in assisting the public in coping with such issues. However, those librarians must have a good understanding of how to navigate the Web given these issues. Yet, many reference librarians feel uncomfortable in dealing with government information; to them navigation of government information resembles having to cope with a "foreign language," one for which they have received inadequate training. Even students in graduate programs in library and information science tend to avoid a course on government information.

Given the efforts of the national government to advance e-government, librarians should confront their reluctance and biases, and assume a major role in providing their constituent groups with knowledge about how to gain access to government information and services, and to participate in e-governance. The challenges are manifold, but there are numerous advantages given the fact that so many people now use e-government for one purpose or another, and the government is expanding the list of constituents it is trying to serve online. Most importantly, helping the communities they serve to participate fully and effectively in e-government falls within the scope of the missions that most academic and public libraries, as well as their parent organizations, expound.

References

Atlas, R. (2003). Science publishing in the age of terrorism. *Academe* **89**, 15.
Benton Foundation (2003). *Achieving E-Government for All. Highlights from a National Survey.* Working document prepared by D.M. West, Brown University. Benton Foundation, Washington, DC. Available: http://www.benton.org/publibrary/egov/access2003.html, retrieved September 19, 2004.
Brademus, J. (1971). Congress in the year 2000, In *The Future of the U.S. Government: Toward the Year 2000.* (H. S., Perloff ed.) George Braziller, New York, NY.
Capron, W. M. (1971). The Executive Branch in the year 2000, In *The Future of the U.S. Government: Toward the Year 2000.* (H. S., Perloff ed.) George Braziller, New York, NY.
Carlitz, R. D., and Gunn, R. W. (2002). Online rulemaking: A step toward e-governance. *Government Information Quarterly* **19**, 389–405.
Congress. House, Office of the Legislative Counsel. Drafting legislation using XML at the U.S. House of Representatives. Available: http://xml.house.gov/drafting.htm, retrieved September 17, 2004.
Cullen, R., and Houghton, C. (2000). Democracy online: An assessment of New Zealand government web sites. *Government Information Quarterly* **17**, 243–267.
Daukantas, P. (2003a, August 25). Administration will review Section 508 compliance. *Government Computer News*, 38.
Daukantas, P. (2003b, September 22). E-gov sites score high on user satisfaction survey. *Government Computer News*, 68.
Daukantas, P. (2003c, August.4). It's all within two clicks on new BEA site. *Government Computer News*, 14.
Educational Resources Information Center (2003). Draft statement of work. Available: http://www.eps.gov/spg/ED/OCFP/CPO/Reference-Number-ERIC2003/Attachments.html, Task 4.2.1, retrieved May 1, 2003.
Forman, M.A. (2001). Associate Director for IT and E-Government, Achieving the Vision of E-government, Office of Management and Budget, Washington, DC (October 1, 2001), p. 4, available: http://www.hpcc.gov/pitac/pitac-25sep011format.pdf.
Frank, D. (2003, May 19). Online feng shui. *Federal Computer Week* **17**, 40–41.
French, M. (2003, November 3). DOD webmasters learn new lessons. *Federal Computer Week* **17**, 9.
General Accounting Office (2003a). *Electronic Rulemaking: Efforts to Facilitate Public Participation can be Improved,* GA0-03-901, Washington, DC, September.
General Accounting Office (2003b). Electronic Government: Progress and Challenges in Implementing the Office of Personnel Management's Initiatives, testimony of Linda D. Koontz, Director, Information Management Issues, before the House Committee on Government Reform, Subcommittee on Technology, Information Policy, Intergovernmental Relations and the Census, General Accounting Office, Washington, DC.
Government proposal may curtail access to data. (2003). *Academe* **89**, 8.

Hernon, P. (1994). Information life cycle: its place in the management of U.S. government information resources. *Government Information Quarterly* **11**, 143–170.
Hernon, P., Dugan, R. E., and Shuler, J. A. (2003). *U.S. Government on the web*. Libraries Unlimited, Westport, CT.
Halchin, L. E. (2002). Electronic government in the age of terrorism. *Government Information Quarterly* **19**, 243–254.
Hardy, M. (2003, May 19). NASA simplifies access to web resources. *Federal Computer Week* **17**, 30.
Hasson J., and Holmes, A. (2003, September 1). Who we believe. *Federal Computer Week* **17**, 18–20, 22–25.
Hernon, P., Relyea, H. C., Dugan, R. E., and Cheverie, J. F. (2002). *United States Government Information: Policies and Sources*. Libraries Unlimited, Westport, CT.
Hernon, P., Shuler, J., and Dugan, R.E. (1999). *U.S. Government on the Web: Getting the Information You Need*. Libraries, Unlimited Englewood, CO.
Implementing the President's Management Agenda for E-Government: E-Government Strategy (2003). Office of Management and Budget, Washington, DC.
Lisagor, M. (2003, October 13). NASA web site readies for next Mars landing. *Federal Computer Week* **17**, 36–37.
Marchionini, G., and Mu, X. (2003). User studies informing e-table interfaces. *Information Processing & Management* **39**, 561–579.
McClure, C. R., Bishop, A. P., Doty, P., and Rosenbaum, H. (1991). *The National Research and Education Network (NREN): Research and Policy Perspectives*. Ablex, Norwood, NJ.
McClure, C. R., and Sprehe, J. T. (2001). Using U.S. information policies to evaluate federal web sites, In *Evaluating Networked Information Services: Techniques, Policy, and Issues*. (C. R., McClure and J. C., Bertot eds.). Information Today, Medford, NJ.
McKenna, E. (2003, June 9). Gov portals get to work. *Federal Computer Week* **17**, 32.
Michael, S. (2003). OMB Issues Privacy Guidance. *Federal Computer Week* **17**, 11 October 6.
National Aeronautics and Space Administration. *Privacy, Accessibility & Translation Capabilities* (NASA Web Site Privacy Notice). Available http://www.nasa.gov/about/highlights/HP_Privacy.html, retrieved September 19, 2004.
National Digital Information Infrastructure and Preservation Program, http://www.digitalpreservation.gov/.
National Research Council (2003). *Building an Electronic Records Archive at the National Archives and Records Administration: Recommendations for Initial Development*. The National Academic Press, Washington, DC.
Office of Management and Budget. (2000, June 22). Memorandum M-00-13. Available: http://www.whitehouse.gov/omb/memoranda/m00-13.html, retrieved September 19, 2004.
Office of Special Counsel scrubs website. (2004, February 23). *OMB Watcher* 5. http://www.ombwatch.org/article/articleview/2060/1/208, retrieved March 9, 2004.
Presidential Memo: The Importance of E-Government, Egov: The Official Web Site of the President's E-Government, http://www.whitehouse.gov/omb/egov/pres_memo.htm., retrieved May 29, 2003.
Perritt, Jr., H. H. (1995). *Electronic dockets: Use of information technology in rulemaking and adjudication*. Online Report to the Administrative Conference of the United States. Available: http://www.kentlaw.edu/classes/rstaudt/internetlaw/casebook/electronic_dockets.htm, retrieved September 18, 2004.

Pew Internet & American Life Project (2002). *The Rise of the E-Citizen: How People Use Government Agencies' Web Sites*, prepared by E. Larsen and L. Rainie. Pew Internet & American Life Project, Washington, DC.
Relyea, H. C. (2003). E-Gov comes to the federal government, In *U.S. Government on the Web: Getting the Information You Need*, 3rd ed., (P., Hernon, R. E., Dugan, and J. A., Shuler eds.). Libraries Unlimited, Westport, CT.
Reed, M. A. T. (2003a, August 25). GPO info gets saved. *Federal Computer Week* **17**, 47.
Reed, M. A. T. (2003b, June 30). SBA: Web simplicity is a virtue, *Federal Computer Week* **17**, 36.
Reed, M. A. T. (2003c, October 11). Section 508: Agencies still on the learning curve. *Federal Computer Week* **17**, 21.
Report bashes U.K. government web sites. (2003). *The Information Management Journal* **37**, 13.
Sprehe, J. T. (2003, September). Practice makes perfect. *Federal Computer Week* **17**, 40.
U.S. Mint. *Privacy Policy*. Available http://www.usmint.gov/policy/index.cfm?action = full#cookies, retrieved May 15, 2004.
Vest, C. M. (2003). Balancing security and openness in research and education. *Academe* **89**, 23.
Willemssen, J.C. (2003). Managing Director, Information Technology Issues, Electronic Government: Success of the Office of Management and Budget's 25 Initiatives depends on effective management and oversight. Testimony before the Subcommittee on Technology, Information Policy, Intergovernmental Relations and the Census, House Committee on Government Reform, GAO-03-495T. General Accounting Office, Washington, DC, p. 4.

Information Seekers' Perspectives of Libraries and Librarians

Eileen Abels
College of Information Studies, University of Maryland, College Park, MD 20742, USA

I. Introduction

This chapter explores the role of libraries and librarians from the perspective of the information seeker in general and from business school students in particular. In a recent article in *First Monday*, Keller *et al.* (2003) pose the question: "What is a library anymore, anyway?" The answer to this question would be "That depends." It depends upon who you are asking and the perspective from which you are answering the question. The notion of perspective has been raised before in the library and information science literature. Zweizig (1976) noted that users were the focus of studies, they were examined from the perspective of "the user in the life of the library" rather than from the perspective of "the library in the life of the user." More recently, Lipow (1999) noted that librarians discuss how to serve "remote users" when in fact it is the library that is remote to the user.

The term "information seeker" was selected purposefully for this chapter to encompass the broadest perspective possible. Looking at "users," "clients," or "patrons," limits the perspective to those individuals who have incorporated libraries and librarians into their information seeking process. This is a library-centric view, seeing the world through the eyes of the librarian. As will become evident in this chapter, focusing on library users would greatly reduce the scope.

An assumption of this chapter is that all individuals are information seekers and a much smaller subset of individuals are users of libraries and library resources. An even smaller subset of individuals consults librarians or information professionals. For this reason, librarians instead must broaden their perspective and consider how they can support information seekers.

The library needs to become an integral channel of information seeking and the librarian a commonly accepted interpersonal source of information.

Another assumption of this chapter is that there is not one perception of libraries and librarians applicable to all information seekers, but rather different perceptions or mental models depending upon the group of individuals. The notion that different groups of information seekers have different perspectives of libraries and librarians is supported by data from studies of information needs. Factors that influence information needs and information seeking behaviors have been reported in user studies that can be found in the library and information science literature.[1] Personal variables that have been identified as influencing information needs and information seeking behavior include discipline, age, task, rank, gender, institutional setting, and education. This means that academic, public, and special libraries and librarians will be perceived differently. There will be some commonalities within settings, so that to some degree, public librarians will be perceived in the same way. Yet, within each setting, not all libraries and librarians will be perceived in the same way. Culture, both geographic and institutional, will influence the perception of information seekers and that will vary from organization to organization within similar settings. Even within a specific organization or institution, perceptions will vary with specific homogenous sub-groups. For example, in an academic setting, the perspective of undergraduate students will differ from that of graduate students that in turn will differ from the perspective of faculty.

This chapter begins with a discussion of two models of information seeking from the library and information science literature that portray the role of libraries and librarians in information seeking. Then data and research findings that shed light on how information seeking behavior has been influenced are presented. Results of a survey conducted at the Robert H. Smith School of Business at the University of Maryland at College Park provide further evidence of how MBA students there perceive librarians. Finally, an updated model of the typical information seeker's perspective of libraries and librarians is proposed.

II. Background

Two information seeking models from the 1960s and 1970s depict the role of libraries from the user's perspective. Taylor (1968, p. 81) presented his model of information seeking to reflect two communication functions of special

---

[1] See the Annual Review of Information Science and Technology (1966) which has had at least a dozen chapters that report results of user studies.

libraries and information centers: self-help and intermediated help. In Taylor's model, the information seeker first consults personal files and then decides whether to discuss the information need with a colleague or to visit a library or information center. Those who first consult with a colleague may later visit an information center. Those information seekers who do choose to visit the library or information center may search by him/herself or ask for the assistance of a librarian or information professional. In a sense, the process is like a funnel, with all information seekers beginning with a search of personal files, and fewer and fewer information seekers passing from one stage to the next. So, some information seekers will end the process with the search of personal files, others will end the process after communicating with a colleague. Still fewer will visit a library and even fewer will consult with a librarian.

Zweizig (1976, pp. 50, 51) presented a model that examines "the library in the life of the user." In this model, Zweizig includes interpersonal sources of information, media, government agencies, and public libraries. Unlike the Taylor model that focused on special libraries, Zweizig notes that the relationship among the various sources of information has not been examined.

The world today is a different place than it was in 1968 and in 1977. Information technology has developed at a rapid pace and has had a huge impact on access to information and on information seeking behaviors. Technology has also influenced lifestyles in general and these changes in lifestyle, specifically the "24 × 7" (24 hours per day, 7 days per week) expectation, have had repercussions on information access and information seeking. Many of these trends are described in the 2003 OCLC Environmental Scan (De Rosa *et al.*, 2004).

A. Personal Factors and Information Seeking

Personal factors identified in studies, such as age and education continue to influence information seeking behaviors. Wiegand (1998) proposed the notion of "personal information economy." By this, he means that different types of information hold a different value to different people because of personal values which are influenced by many factors. In addition to differences based on personal values and factors influencing use, there are factors that cause information access gaps. According to the Pew Internet and American Life Project, Internet access has grown across the board but there are clear demographic gaps by age, income, ethnicity, educational level, and geographic location (Lenhart *et al.*, 2003, p. 4).

Other variables that were identified as influencing information seeking behaviors are perceived ease of use, perceived accessibility, and prior experience (Allen and Gerstberger, 1967; Rosenberg, 1967). These factors, identified as having played a role in the adoption of the Internet (Abels *et al.*, 1996), most likely continue to influence information seeking behaviors. In general, information seekers select sources of information that are easy to use and easily accessible, even if the quality of the information provided was not considered to be as high. Channels of information and sources of information that have barriers to use will not become a part of information seeking behaviors. Library services are fraught with barriers. Fagan and Ruppel (2002) found that students perceived many barriers to asking questions at the reference desk. Accessing library resources from outside the library presents various barriers, unlike accessing free resources on the Web. Due to licensing constraints, the number of simultaneous searches may be limited, user validation is required, search capabilities may be limited, and some library resources simply may be available only to those who visit the library. Simplifying access to resources and eliminating barriers would likely increase usage.

B. Demographics

Information seekers today represent four distinct generations (with a fifth generation about to begin to seek information). Zemke *et al.* (1999) discuss differences in the four generations from the perspective of training. Many characteristics of information seeking relate to characteristics of training. There is no one size fits all and each generation has its own style. For training, the authors recommend understanding the sociology of the four generations, offering many options, using their icons, language and precepts to "accommodate personal scheduling needs, work–life balance issues and learning styles (p. 54)." Table I synthesizes and summarizes characteristics that differentiate the information seeking styles of the different generations (Zemke *et al.*, 1999, 2001; Lancaster, 2003).

C. Technology Beyond the Internet

Laptops, personal data assistants (PDAs), and cell phones now define our society. PDAs and cell phones offer text messaging capabilities. Wireless access is being offered in more and more places, including coffee shops and fast-food restaurants (Rose, 2003). Instant messaging which gained huge popularity among pre-teens and teens has to a great extent replaced the telephone and to some extent e-mail. Instant messaging now allows

Table I
Information Seeking Behavior and the Generations

Generation	Traditionals (Silents, Veterans)	Boomers	X-ers	Millenials (Nexters)
Approximate dates of birth	1922–1943 (pre-WWII)	1943–1960	1961–1980	1981–1999
Characteristics that influence information seeking	(a) Accustomed to top–down flow of information (b) Formal (c) Learning environment that is stable	(a) Formal feedback (b) Interactive and non-authoritarian	(a) Highly independent (b) Entrepreneur (c) Comfortable with change (d) Raised with instant access (e) Want frequent, immediate feedback (f) Self-directed (g) Sample and learn by doing (h) Not attracted to classroom	(a) Globally concerned (b) Diverse (c) Cyberliterate (d) Media savvy (e) Collaborative (f) Multitaskers (g) Teamwork (h) Technology (i) Multi-media
Information seeking	(a) Like materials that are organized and summarized (b) Example: Reader's Digest (c) Dewey Decimal	(a) Easy to scan format (b) Example: Business Week; USA Today; People	(a) Prefer fewer words (b) Do not read as much (c) Visual stimulation—headlines, subheads, quotes, graphics, lists (d) Example: Spin, Fast Company, Wired, chat-room dialogue	(a) Readers (b) Lively and varied materials (c) Chat (instant messaging) (d) Search engine (Google)

Source: Summarized from Zemke *et al*. (1999, 2001), and Lancaster (2003).

the addition of video and voice. Instant messaging has expanded beyond the home and now finds itself in the workplace. Information seekers find themselves faced with a wide array of channels to access information, libraries and librarians. They have access any time and any place.

D. Self-help and 24 × 7

With access from home on the rise, information seekers now expect to be able to shop, make travel arrangements, send e-mail, and chat with friends any time, or "24 × 7." The increase in the self-service trend is evident in a variety of industries (Harmon, 2003; De Rosa et al., 2004). While self-help has been available in banks and gas stations for some time, there is a trend toward self-service in grocery and other retail stores. Airlines now use automated kiosks for checking in passengers. Some ski resorts in Colorado have installed "skiosks" that dispense lift tickets. The growth of self-service machines is due in part to improved technology. However, Harmon (2003) notes that another reason to turn to self-help machines is to avoid frustrating experiences with service workers. In any case, the trend toward self-help is growing.

III. Information Seeking Behaviors Today

The impact of these technological and lifestyle changes on information seeking behaviors today is profound. Data from various reports and articles indicate that while the medium used may have changed, information seekers still do not select libraries or librarians as their primary channel for information. One clear change from the Taylor model is the addition of the Internet as an information channel. According to the CyberAtlas (2003), as of November 2003, the Internet Universe Estimate (defined as people over the age of 2 that have access to the Internet from a personal computer at home; includes active and non-active persons in the household) was 421,653,760. The report released from the Pew Internet & American Life Project in April 2003 indicated that 24% of Americans are "truly disconnected"; that is, they do not use the Internet directly or have indirect access to the Internet through household members (Lenhart et al., 2003, p. 3). Overall, the data gathered by the Pew Internet Project indicate that Internet penetration in the home has been hovering between 57% and 61% (p. 3) since late 2001. In 2000, Pew reported that 49% of American adults had Internet access. In 2002, the percentage rose to 58%. Interestingly, Crawford (2003, p. 42) notes that daily newspaper readership is at the same level as Internet usage (58%).

Casey (2000) presented data on where people turn for information. While in 1998, more people turned to libraries for information than over

Table II
Where Do People Turn for Information?

	December 1998	June 2000
Internet	22.8	35.7
Books	14.6	12.3
Library	24.3	12.3
Friends	8.4	9.0

Source: Casey (2000).

the Internet, the numbers just 2 years later were astounding: library users had dropped by half and Internet usage had greatly increased. The data are summarized in Table II.

The Association of Research Libraries (ARL) provides statistics on the use of library services that help paint a picture of users' perspectives of libraries and library services (ARL Statistics). As can be seen in Table III, both the median number of reference queries and the median number of total circulation had a steady decline between 1997 and 2002. Yet, expenditures on monographs showed a small steady increase between 1997 and 2001, showing a decline only in 2002. If academic library users are turning to other types of resources, one might expect a decrease in expenditures on monographs. Perhaps the decrease in 2002 monograph expenditures is reflective of a downward trend.

Table III
ARL Statistics on Reference Queries and Circulation

Year	Reference queries	Circulation	Expenditures on monographs ($)
1997	162,336.5	527,993.5	1,453,699
1998	147,644	510,310	1,470,005
1999	129,482	509,655	1,506,650
2000	121,637	476,690	1,645,248
2001	109,713	467,277	1,833,221
2002	100,656	464,704	1,806,964

Source: ARL Statistics Interactive Edition.
Available at: http://fisher.lib.virginia.edu/cgi-local/arlbin/arl.cgi?task=setupstats <Accessed February 7, 2004>.

Interestingly, in some academic libraries that have undergone recent facility renovation, gate counts have increased even while both circulation and reference requests have declined (Shill and Tonner, 2003). It seems that the information seeker's perception of the library may be focused on library as place; the students at universities seek a place to study, work in groups, or use computers. Since the Shill and Tonner study focused on the library and not on the library visitor, the reasons for library visits are conjecture. However, there seems to be a trend in library renovations to include cafes or coffee shops.

Other academic libraries that reported a drop in gate counts and circulation records did report a large increase in the use of electronic resources, which can be accessed remotely (Carlson, 2001). End-user access to resources is not new to librarians. Online services were introduced in the 1960s (Bjorner and Ardito, 2003). However, much remote access was restricted to the librarian and the use of CD-ROMs required users to visit the library for access. We now live in a world where high-speed access is becoming the norm and remote access via the Web is commonplace. Libraries now provide desk-top access to end-users who search from their offices or homes. Electronic collections now include electronic books, electronic journals and a variety of fulltext databases. While personal computers are still popular, laptops are increasingly replacing personal computers for office and home use. Within libraries, laptop usage is increasing as access to wireless networks is becoming more common.

A. Trends in Library Services

Library services are reflecting changes based on new technologies and lifestyle trends. Many libraries now offer self-help services that include checking out, renewing, and reserving material, requesting interlibrary loans, and accessing databases from remote locations. Sackett (2001, p. 209) notes that at the University of Kentucky, "self-service vending machines allow users to buy and add money to a single card to be used in the library's copiers, pay print stations, and even the cafe cash registers."

B. Communicating with Librarians

While the Taylor model implied only face-to-face communication as the norm, technology now provides many more options. Reference services are now offered at reference desks face to face, via telephone, e-mail, and chat. Fagan and Ruppel (2002) summarized several users studies from the 1970s to 1980s that summarized students' perceptions of academic librarians. They enumerated several barriers to asking questions at the reference desk: feeling that the librarian was too busy to interrupt, and not knowing which questions to ask. The OCLC Environmental Scan recognizes barriers to library resource access by posing questions on ways to make libraries and library content more accessible (De Rosa *et al.*, 2004).

Special and academic libraries report heavy usage of e-mail reference; e-mail is ubiquitous; it is heavily used by the members of organizations, students, and faculty. E-mail reference is an extension of that usage. Special libraries shy away from the use of forms for e-mail requests because filling out a form may present a barrier to the information seeker. Sending an e-mail to the library should be as easy as sending an e-mail to a colleague or friend.

Breeding (2003, p. 39) notes that instant messaging "emulates the hallway conversations—where the best exchange of ideas often takes place—much better than e-mail". Instant messaging is now being used for customer service and communicating with colleagues in different geographic locations. Libraries are currently offering live chat services to patrons, many 24 hours a day, 7 days a week (Jackson, 2002). Breeding suggests that the success of virtual reference hinges in part on the comfort of librarians in the chat environment. He notes that while libraries are attempting to offer virtual reference services using chat technology, in fact, libraries have been slow to incorporate instant messaging for internal business communication. As Breeding suggests, usage will likely increase this comfort level. Furthermore, as was noted above in the discussion of different generations, the comfort level with chat varies greatly. Generation X-ers and Millenials are more comfortable in the live chat arena and multi-tasking does not daunt them.

The means with which librarians have incorporated chat into virtual reference services, generally requires a person to visit the library Web site. While the librarian may perceive this as a trivial complaint and consider access via the library Web site as highly accessible, information seekers might not agree. Switching to a Web site or even a page not currently in use can be a barrier to usage; having the library added to "buddy lists" on instant messaging services being used by the information seeker could increase the use of this service a great deal. In fact, it could increase the usage beyond capacity. Overall usage of chat services varies greatly yet the numbers are not overwhelming at this point. According to Coffman (2003) virtual reference services receive an average of less than 10 questions per day.

Horn (2001) noted that the librarians' perceptions of user needs and expectations have changed thus prompting librarians to offer digital reference services to provide "users with access to a knowledgeable librarian at the user's convenience rather than just during hours when the reference desk is open (p. 320)." Jackson (2002) questions whether data support adoption of the 24 × 7 and AskJeeves commercial models by librarians.

C. New Library and Librarian Competitors

According to a *Wall Street Journal* article technology is "disrupting basic business models, plunging companies into new markets, creating new competitors and blurring the boundaries between industries" (Angwin *et al.*, 2004). This certainly applies to libraries who now find themselves competing with bookstores, both physical and virtual, the Internet in general and search engines specifically.

Goodman (2003) quotes a high school student who notes: "The Internet has become such a major part of doing research." Information seekers use a variety of tools and resources on the Internet, with search engines being the number one tool. According to a survey conducted by iProspect, "77% of Internet users employ search engines more frequently than any other online media (iProspect, 2002)." Sullivan (2003) reported the number of searches performed per day on eight search engines in February 2003: on the low end of the range FAST reported 12 million searches per day while Google was on the high end with 250 million searches per day.

Many digital collections and resources are available to individual subscribers as well as to institutional subscribers. This allows information seekers to access resources without the library as intermediary. Other resources offer some content for free and more complete content for a fee. *Encyclopedia Britannica* (www.britannica.com) is one example of a resource offering some content for free.

Still another model is to target only individual researchers, not institutions. HighBeam™ Research (formerly e-library.com, now located at http://www.highbeam.com/Library/index.asp) provides access to approximately 28 million documents including newspapers, magazines, transcripts, books, images, photographs, and reference works. Questia.com, claiming to be the "word's largest online library," is restricted to individual subscriptions because of licensing agreements, essentially removing libraries from the loop. On its Web site, Questia.com recognizes the needs of various user groups—students, teachers, librarians, and publishers. Furthermore, this "library" has enhancements; researcher may highlight text, take notes, and generate bibliographies. Questia.com attracts 1.2 million visitors a month (Goodman, 2003). Comparing Questia.com to public libraries, Goodman (2003) notes,

> while many public libraries provide roughly the same services for free, these subscription sites often cover a wider ranges of subjects and offer more copyrighted materials, all within the confines of a kid's home.

IV. MBA Students' Perceptions of Librarians and Libraries: A Case Study

The above discussion drew on generalities about information seekers, but as was noted, not all information seekers will perceive libraries and librarians in the same way. Business school students were targeted early on as a group of information seekers highly dependent on the Internet. Bell (1998) noted that MBA students were "webcentric," using only Internet sources for business research. Morrison and Kim (1998) reported similar findings: business school students used free Web resources more than any other type of resource.

There is a great deal of LIS literature on user education and the role of the academic librarian as instructor (Dewald, 1999; Kilcullen, 1998; Patterson, 1987). Several articles focus specifically on workshops and user education efforts for business school communities (Flanagan, 1999; Gunderson, 1991; Judd and Tims, 1996). In general, it appeared that workshops are poorly attended. Gunderson noted that business school students asked, "Why do I have to waste my time with this class?" (p. 31) in reaction to a one-credit course offered by the Library and the College of Business and Administration at the University of Colorado.

Business school students may also represent information seekers in the business world as well. Katzer and Fletcher (1992) reported that managers spend most of their time communicating with others, thus favoring oral channels. These findings echoed reports by Mintzberg (1989) that managers spend between 66% and 80% of their time in oral communication. Auster and Choo (1993) noted that libraries and electronic information services were not frequently used in environmental scanning activities by chief executive officers. These findings suggest that libraries and librarians do not play a central role in information seeking in the business environment.

In an effort to gain insight about the information seeking behaviors of business school students at the University of Maryland at College Park (UMD), an electronic survey was administered to students in the MBA program at the Robert H. Smith School of Business (RHS) during the spring semester 2002. At the time, the student population consisted of 450 fulltime MBA and 663 part-time MBA. As is the case with many Web surveys, there was no control maintained to track respondents. Demographic questions on status and length of time at RHS were used to track the representativeness of the sample. E-mail messages were sent to all full-time and part-time MBA and MS students with a link to the survey. The survey was available for a 2-week period; a reminder e-mail message was sent after 1 week. In order to enhance the response rate, respondents were entered into a drawing for small prizes.

The survey consisted of 23 questions divided into four parts. Part 1 consisted of seven questions about information seeking behaviors and perceptions about information seeking using a seven-point Likert scale. Part 2 consisted of eight questions about information seeking and information needs. Part 3 consisted of one open-ended questions about ways in which librarians have assisted or could assist in business research. Part 4 consisted of six multiple choice questions relating to demographics and logistics as well as one final open-ended question. Only those questions relating to users perspectives of the role of the librarian and the library will be discussed here.

At the end of the survey period, 243 usable responses were received out of the possible 1163, giving a response rate of 21%. The number of responses from fulltime (112) and part-time (118) students were not entirely representative of the actual enrollment at the time with a higher response from the part-time students. More than half of the respondents (57%) were first year students. While a limitation to the study is the low response rate, the data are considered to represent general patterns of information seeking; in the worst case scenario, one would assume that all non-respondents are non-library users and that this sample represents a higher percentage of library users and of potential library users than is actually the case.

Although consultation with a librarian was higher than expected, findings indicate that this group of business school students is like other user groups in that consultation with librarians is relatively low. The survey was designed to solicit responses to this question in two ways. While 40% of the respondents agreed or strongly agreed that they have consulted a librarian at UMD or elsewhere when conducting business research, only 20% of the respondents indicated that they are likely to seek assistance from a librarian in general when conducting business research. Only 15% of the respondents indicated that they tend to visit a library in person when conducting business research.

The survey sought to understand the role of different sources used when conducting business research projects. Respondents were asked how they would most likely begin the business research process, restricting responses to those who indicated they would consult a librarian when conducting business research. Table IV summarizes the results.

To delve deeper into the perception of librarians held by business school students, the role of librarian in the overall research process was explored. Sixteen individuals, 17% of those respondents who have not consulted a librarian for business research ($n = 95$), indicated that they would consider consulting a librarian. This means that 7% of the total respondents ($n = 243$) fall into the category of potential users of an intermediated service.

Table IV
Sources Most Likely Consulted when Beginning Business Research

Source	Percentage ($n = 160$)
Web using a search engine	69
Virtual Business Information Center (VBIC)[a]	17
Favorite database available through subscription	10
Librarian	3
Other	1

[a] VBIC is a tool developed by a team from the College of Information Studies, the UMD Libraries, and RHS.

Respondents were asked about the type of assistance that they would like to receive from librarians. Responses included the following:

- 78% ($n = 157$) of the respondents would ask for assistance in selecting resources;
- 78% ($n = 157$) of the respondents would ask for assistance using a database available through the library;
- 61% ($n = 157$) of the respondents would ask for assistance in developing a search strategy;
- 53% ($n = 241$) of the respondents agreed or strongly agreed that they would use a Web site created by librarians.

Interestingly, more than half of the respondents (55%, $n = 157$) indicated that they would prefer to learn how to find the information for themselves when consulting a librarian. This finding reflects the self-help trend discussed earlier in the chapter. Yet, 22% of the respondents would prefer to receive an answer to the question posed, supporting the notion that there are multiple patterns to information seeking, even within a relatively homogeneous group.

A related matter is the preferred means of communicating with librarians for those respondents who would communicate with a librarian. In general, 41% of respondents ($n = 241$) agreed or strongly agreed that they would definitely consult a librarian via a virtual reference desk. Table V, which is limited to those respondents who have or would consider consulting a librarian ($n = 160$), shows respondents' reactions to different methods of communicating with librarians. While all means of contact might be used, more than half of the respondents indicated that they might use two means of communication with a librarian: the virtual reference desk and e-mail. Only 7% would prefer to talk with a librarian at the UMD Library and 13% would

Table V
Business School User Preferences for Means of Contacting Librarians

Means of contacting the librarian	Might use ($n = 161$) (select all that apply)	Preference to use ($n = 161$) (select one)
Use a virtual reference desk	68%	21%
Send an e-mail message	65%	16%
Talk in person at the UMD library	47%	7%
Talk in person at the business school	46%	13%
Talk on the telephone	47%	10%
Would not consult a librarian	N.A.	33%

prefer to talk in person with a librarian at the business school. These results are not surprising. The preference for electronic communication seems to fit with the overall trends in electronic communication in general and among business school students. However, to serve the broadest range of business school students, librarians would have to be available from all communication channels to some extent.

The survey also sought feedback on the level of interest in workshops conducted by librarians. Approximately 70% of the respondents ($n = 241$) indicated an interest in workshops related directly to course assignments; 64% of the respondents indicated an interest in workshops on specific business topics; 61% indicated an interest in workshops that focus on specific databases.

V. Discussion

The survey data and trends presented in the literature review suggest that librarians and libraries are not perceived as the first stop in information seeking. However, librarians may be perceived as an instructor at least among certain information seekers. Information seekers who want to be self-sufficient in their searching may seek assistance in accomplishing that. Table VI shows ARL statistics on the median number of group sessions and the median number of session participants over a 5-year period. The median number of sessions has fluctuated, with the median number of sessions in 2002 showing an increase of 8% over 1997. The median number of participants has shown a steady increase between 1997 and 2002. The median number of session participants has increased approximately 25% since 1997.

Table VI
ARL Statistics on User Education

	Median number of group sessions	Median number of participants
1997	700	9,311
1998	736	9,786
1999	714	9,585
2000	721	9,799
2001	696	10,657
2002	757	11,712

Source: ARL Statistics Interactive Edition.
Available at: http://fisher.lib.virginia.edu/cgi-local/arlbin/arl.cgi?task=setupstats <Accessed February 7, 2004>.

However, to reach a broader audience given the usage of electronic communication channels, perhaps librarians should pursue alternative approaches to in-person user instruction sessions. Flanagan (1999) captured this idea when she created an instructional tool that could be used "any time, any where." However, librarians would do well to heed Stoan's (1991) warning that library instruction tends to ignore the "broader intellectual and social context in which scholars function as information-generating and information seeking individuals (p. 238)." In other words, librarians should create instruction opportunities that fit into the information seekers communication patterns. Table I, which focused on the differences in information seeking behaviors among members of different generations, should be taken into consideration in developing methods of instruction.

Based on the literature review and the data presented above, what can we conclude about how information seekers in general and business school students in particular perceive librarians and libraries? While Taylor's and Zweizig's models of information seeking are still valid to some extent, they are in need of updating to reflect the various technological, lifestyle and demographic trends. Rather than consider the user, this model considers the information seeker in his or her environment. Figure 1 presents a proposed updated model of information seeking from the perspective of an information seeker that reflects technological advances and combines sources and channels from both the Taylor and Zweizig model as well as additional sources and channels.

The updated model incorporates information received by the information seeker from the media as well as various agencies and institutions. Unsolicited information received from the media and agencies influences

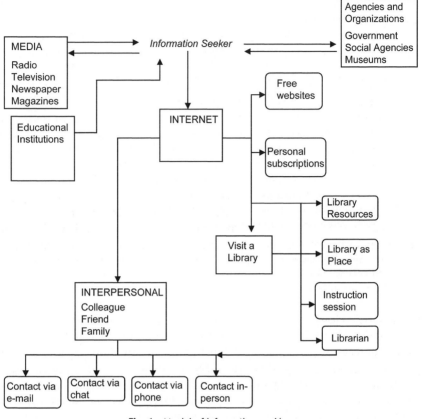

Fig. 1 Model of information seeking.

overall information seeking behavior. In addition, the information seeker may turn to the media for information, for example, to learn about election results, or news regarding world events.

In the updated model, the notion of personal files has greatly expanded. In the 1960s and 1970s, the information seeker would have checked the file cabinets and print books and journals on nearby bookcases. Today, information seekers check both print and electronic files, including documents stored on personal computers, laptops and other electronic devices. Technology now allows information seekers to search beyond their own personal files without leaving their offices or homes; information seekers search the World Wide Web for information. Thus, an information seeker may access resources near or far with a computer and network connection close at hand. The updated model assumes that the first stop in information seeking is the Internet.

In the 1960s and 1970s, consultation with colleagues primarily would have occurred face to face or perhaps by telephone. Now, of course, communication may also occur via electronic mail and chat. It seems that the librarian is not considered among the first interpersonal sources to be consulted and for that reason, in this model, the librarian is placed within the confines of the library. The proposed updated model places interpersonal sources after consultation of the Internet, however, the role of interpersonal sources is not clear and further research is needed to validate the model and to determine where interpersonal sources fit into the information seeking process.

Consulting the library formerly required a trip to the library or, perhaps, a telephone call to a librarian. As was the case with the Taylor model, a larger group of information seekers will visit a library, either in-person or virtually, than will consult with a librarian. Consultation with librarians does not require an in-person visit or even a telephone call; information seekers may send an e-mail message to a librarian or in some cases, may chat real-time with a librarian.

VI. Conclusions

Different groups of information seekers have different perspectives of libraries and librarians. Librarians cannot impose their perspective upon the information seeker. Moving away from the notion of user/non-user to information seeker moves a library-centric perspective to the perspective of the information seeker.

Research that has been conducted to date suggests that:

- The Internet has become central to information seeking by many groups of information seekers.
- Libraries and librarians are not central to information seeking in general.
- Academic library "users" view the library as place and the librarian as instructor.
- Factors, such as ease of use and accessibility, continue to influence information seeking.

Librarians must continue to monitor technology and lifestyle changes. Whatever technology libraries and librarians adopt should fit the culture of the organization and the information seeking behavior of those in the setting. Adoption of technology should be based on evidence that supports adoption; evidence that validates the information seekers' perspective. Further research is needed to fully develop and validate the current model of information seeking and to determine differences in the model based on different demographic groups. Most importantly, the librarians need to accept the broader framework of the information seeker and develop services that integrate the library and the librarian into this framework.

References

Abels, E. G., Liebscher, P., and Denman, D. W. (1996). Factors that influence the use of electronic networks by science and engineering faculty at small institutions. Part I. Queries. *Journal of the American Society for Information Science* **47**(2), 146–158.
Allen, T. J., and Gerstberger, P. G. (1967). *Criteria for Selection of an Information Source.* (Working Paper No. 284-67). Alfred P. Sloan School of Management, Massachusetts Institute of Technology, Cambridge, MA.
Angwin, J., Peers, M., and Squeo, A. M. (2004). Drive force behind offer: technology tears up old business models. *Wall Street Journal*, 12 February 2004, 1A.
Annual Review of Information Science and Technology (1966). Learned Information, Inc., Medford, NJ. v. 1.
ARL Statistics Interactive Edition. (On-line), Available at: http://fisher.lib.virginia.edu/cgi-local/arlbin/arl.cgi?task = setupstats (accessed February 7, 2004).
Auster, E., and Choo, C. W. (1993). Environmental scanning by CEOs in two industries. *Journal of the American Society for Information Science* **44**(4), 194–203.
Bell, S. J. (1998). Weaning them from the Web: teaching online to the MBA Internet generation. *Database* **21**(3), 67.
Bjorner, S., and Ardito, S. C. (2003). Online before the Internet: early pioneers tell their stories. *Searcher* **11**(6), 36–46.
Breeding, M. (2003). Instant messaging: it's not just for kids anymore. *Computer in Libraries* **23**(10), 38–40.
Carlson, S. (2001). The deserted library. *Chronicle of Higher Education* **48**(12), A35.
Casey, T. (2000). *Where People Turn for Information.* Paper presented at the Association for Library Trustees and Advocates Program, American Library Association Conference, Chicago, IL, July 9, 2000.
Coffman, S. (2003). *Going Live.* American Library Association, Chicago.
Crawford, W. (2003). Fleeing the Internet: time to call it quits? *Econtent* **26**(11), 42–43.
CyberAtlas (2003). *Global Usage.* (On-line), Available at: www.cyberatlas.internet.com (accessed 23 January 2004).
De Rosa, C., Dempsey, L., and Wilson, A. (2004). The 2003 OCLC Environmental Scan: Pattern Recognition: A Report to the OCLC Membership. OCLC, Dublin, OH. (On-line) Available at: http://www.oclc.org/membership/escan/toc.htm (accessed 26 January 2004).
Dewald, N. W. (1999). Web-based library instruction: what is good pedagogy? *Information Technology and Libraries* **18**(1), 26–31.
Fagan, J. C., and Ruppel, M. (2002). Instant messaging reference: users' evaluation of library chat. *Reference Services Review* **30**(3), 183–197.
Flanagan, D. L. (1999). Learning anytime, anywhere: designing Web-based business tutorials. *Journal of Business & Finance Librarianship* **4**(4), 19–31.
Goodman, L. M. (2003). E-Commerce (A special report): Consumer guide – writing tools: for students researching a paper, online libraries are increasingly the way to go; Here's how they stack up. *Wall Street Journal*, 16 June 2003, R11.

Gunderson, L. (1991). Teaching business sources to MBA students. *Colorado Libraries* **17**, 31–32.
Harmon, A. (2003). More consumers reach out to touch the screen. *New York Times*, 17 November 2003, Section 1, Page 1, Column 2.
Horn, J. (2001). The Future is now: reference service for the electronic era. In: *Crossing the Divide: Proceedings of the Tenth National Conference of the Association of College and Research Libraries*, 15–18 March 2001, Denver, CO. Available at: http://www.ala.org/ala/acrl/acrlevents/horn.pdf (accessed 20 April 2004).
iProspect survey finds search engines beat all other media for driving visitors to Web sites with 77% of Internet users employing search engines to find Web sites. *PR Newswire*, 16 September 2002, (On-line). Available at: http://www.findarticles.com/cf_dls/m4PRN/2002_Sept_16/91531798/p1/article.jhtml (accessed 23 January 2004).
Jackson, M. (2002). A rush to serve: digital reference services and the commitment to 24/7. *Advances in Librarianship* **26**, 299–317.
Judd, V. C., and Tims, B. J. (1996). Integrating bibliographic instruction into a marketing curriculum: a hands-on workshop approach using interactive team-teaching. *Reference Services Review* **24**(1), 21–30 and 56.
Katzer, J., and Fletcher, P. (1992). The information environment of managers. *Annual Review of Information Science and Technology* **27**, 227–263. Learned Information, Inc. Medford, NJ.
Keller, M. A., Reich, V. A., and Herkovic, A. C. (2003). What is a library anymore, anyway? *First Monday*, (On-line). Available at: http://firstmonday.org/isseus/issue8_5/keller/ (accessed 23 January 2004).
Kilcullen, M. (1998). Teaching librarians to teach: recommendations on what we need to know. *Reference Services Review* **26**(2), 7–18.
Lancaster, L. C. (2003). The click and clash of generations. *Library Journal*, 36–39.
Lenhart, A. *et al.* (2003). The ever-shifting Internet population: a New look at Internet access and the digital divide. Pew Internet & American Life Project, (On-line). Available at: http://www.pewinternet.org/reports/toc.asp?Report = 88 (accessed 23 January 2004).
Lipow, A. G. (1999). Serving the remote user: reference service in the digital environment. Keynote address at the Ninth Australasian Information Online and On Disc Conference and Exhibition in Sydney, Australia, January 1999. Available at: http://www.csu.edu.au/special/online99/proceedings99/200.htm/ (accessed 20 April 2004).
Mintzberg, H. (1989). *Mintzberg on Management*. Free Press, New York.
Morrison, J. L., and Kim, H. (1998). Student preference for cybersearch strategies: Impact on critical evaluation of sources. *Journal of Education for Business* **73**(5), 264–268.
Patterson, C. D. (1987). Librarian as teachers: a component of the educational process. *Journal of Education for Library and Information Science.* **28**(1), 2–8.
Rose, K. (2003). Network, rest and play: Offering your customers wireless Internet access may mean that they spend more time, and money, in your venue. *Technolgy – WiFi Leisure Report* **18**(1), 1–8.
Rosenberg, V. (1967). Factors affecting the preference of industrial personnel for information gathering methods. *Information Storage and Retrieval* **3**, 119–127.
Sackett, J. H. (2001). Planning the new central academic library. *New Library World* **102**(6), 207.

Shill, H. B., and Tonner, S. (2003). Creating a better place: physical improvements in academic libraries, 1995–2002. *College and Research Libraries* **64**(6), 431–466.

Stoan, S. K. (1991). Research and information retrieval among academic researchers: implications for library instruction. *Library Trends* **39**(3), 238–257.

Sullivan, D. (2003). *Searches Per Day*, 25 Feburary 2003, (On-line). Available at: http://www.searchenginewatch.com/reports/article.php/2156461.

Taylor, R. S. (1968). Question-Negotiation and information seeking in libraries. *College & Research Libraries* **29**, 178–194.

Wiegand, W. (1998). Mom and me: a difference in information values. *American Libraries* **29**(7), 56.

Zemke, R., Raines, C., and Filipczak, C. (1999). Generation gaps in the classroom. *Training* **36**(11), 48–54.

Zemke, R., Raines, C., and Filipczak, B. (2001). Generation markers. *Across the Board* **39**(4), 20.

Zweizig, D. L. (1976). With our eye on the user: needed research for information and referral in the public library. *Drexel Library Quarterly* **12**, 48–58.

Competition or Convergence? Library and Information Science Education at a Critical Crossroad

Joan C. Durrance
Margaret Mann Collegiate Professor of Information, School of Information, University of Michigan, Ann Arbor, MI, USA

I. The Changed Information Landscape

Libraries and librarians have long been early adopters of information technologies. For decades, librarians have applied computerization to library operations. Standardization and computerization of bibliographic records decades ago made possible automation of library systems, the creation and utilization of giant bibliographic utilities such as OCLC with its 52 million records. Collaborative adoption of information technologies decades ago brought shared cataloging, on-line public access catalogs, bibliographic databases, enhanced interlibrary loan and document delivery, and acquisition of information in digital formats, resulting in worldwide access to library resources. Nonetheless the revolution in information technologies that produced the World Wide Web in the mid-1990s hit the information profession of librarianship and the educational establishment like an earthquake.

As librarians vividly recall, the changes that resulted in radically different approaches to access to information content, and the ability to communicate and collaborate around knowledge, brought fear to many in Library and Information Science (LIS). Computer scientists promised that intelligent agents would provide direct access to Internet content and in the process would bypass intermediaries, including librarians. Some feared that the Internet would make both libraries and librarians superfluous, or doom them to extinction. Looking back a mere decade, it is not difficult to see that the Internet and the changing information infrastructure brought at least fear of a crisis. Built by computer scientists, the Internet provided radically new kinds of information flow and created a new information landscape—just as an earthquake might. The continuing information revolution has been at the same time the most serious crisis ever faced by the field and its biggest opportunity.

The changed landscape has influenced both practice and education. Added to this situation is the current shortage of professional staff, so profound that it has been noted by the press, and by the First Lady who has spearheaded a major recruitment effort (Tenopir, 2002; Lynch, 2002). In addition, statistics predict the impending retirement by the end of the next decade of 68% of the nation's librarians (Tenopir, 2004; Lynch, 2002). These factors have focused attention on the education of librarians.

This paper focuses specifically on the changes that are ongoing in the evolution of LIS education; it is framed as a discussion of crisis and opportunity. Continuing change in the information environment is a compounding factor in the crisis; while information technologies have had a profound impact on libraries and librarians, a group of vocal librarians appear to resist the inevitable changes that must occur in educating information professionals for the future. Factors external to the field and the field's changing research paradigm, to be discussed below, have put LIS educators in direct competition with others for the domain that had been claimed since the end of the 19th century by librarians. This paper examines the changed landscape in which LIS education operates and its accompanying opportunities for the domain that LIS claims.

II. It Is Not Just Technology: The Changing Research Paradigm

The knowledge base of LIS, built largely by faculty in LIS programs, grew slowly throughout much of the 20th century, but experienced rapid growth in the last two decades well before the Internet crisis of the 1990s. In the process of the growth of the knowledge base of the field, increasing numbers of researchers realized that "library problems" were actually "information problems." Over 30 years ago Robert Taylor, dean of the Syracuse program from 1972 to 1981, suggested moving from the Ptolemaic and library-centered view of the universe to a "dynamic Copernican universe with information at its center and with libraries playing a significant, but not necessarily central, role." (Sutton, 2001) This paradigm has come to be accepted by LIS researchers and has permitted them to develop more effective frameworks for their work which today encompasses what Marcia Bates, UCLA faculty member, researcher, and theoretician, has summarized as the three "Big Questions" of LIS research: "(1) The physical question: What are the features and laws of the recorded-information universe?

(2) The social question: How do people relate to, seek, and use information? (3) The design question: How can access to recorded information be made most rapid and effective?" (Bates, 1999).

Researchers seeking the answers to the three big questions identified by Bates have moved LIS education into a Copernican universe with information at its center and, as Taylor predicted, libraries as one of the planets in the universe. The knowledge gained from this approach has entered LIS education and has influenced the changes identified in the KALIPER study discussed below. The changed information landscape and the changing LIS research paradigm have been accompanied by intensive internal examination of domain knowledge within LIS and more broadly within the framework of information, rather than the library framework. These changes, of course, have occurred program by program. Broad-spectrum examinations of various scholars looking across this changing field have produced both a variety of thoughtful, penetrating journal articles and a group of serious monographs that seek to synthesize the knowledge domain and contributions of LIS. This section of the chapter examines some of this work.

III. Operating in a Highly Competitive Environment

Eight years ago LIS researchers Nancy A. Van House and Stuart A. Sutton, re-examined the Ptolemaic vs. Copernican debate started in the 1970s by scholars such as Robert Taylor, but made an urgent priority by the advent of the Internet. Using an ecological metaphor which compared library-focused education to the Panda Syndrome, these scholars noted that the panda is nearing extinction because of its limited ecological niche arguing that narrowly focused library education programs were doomed to extinction.

> LIS education is operating in an extremely dynamic and highly competitive environment. The growing importance of information, developments in information technology and the information environment, and LIS' own efforts at adaptive radiation have created an ecological convergence between LIS and other professions and professional education programs both in LIS' traditional niche (e.g., "digital libraries") and new niches (e.g., information management). The information field is undergoing radical change, and LIS is not the only profession seeking to claim jurisdiction. (Van House and Sutton, 1996)

Van House and Sutton warned "that the increasing value of information is bringing other professions into the information field, and changing the boundaries and rules of competition" adding that "both the LIS profession and education for LIS...[are] engaged in a struggle with other professions and academic disciplines both for jurisdiction over the information functions that have traditionally been the problem domain of LIS and of the information functions brought about by changes in technology and society." Van House and Sutton further warned that "to compete, LIS education and the profession have to be more cognizant of their own and their competitors' habitus and the dynamics of this changing, enlarged field..."

Noting an ecology of professions, they indicate that "professions are created, grow, transmute, and disappear... Convergence is especially likely when the rewards (money, prestige, power) of a problem area or professional niche are great, attracting attention from many professions" and, quoting Andrew Abbot, they indicate that "knowledge is the currency of the competition. (Abbott, 1988, p. 102)"

While the knowledge gains resulting from LIS research have served to position LIS education for the information technology revolution, they have also served to distance it from practice, creating what some in the field see as a crisis in "library" education (American Library Association. Congress on Professional Education, 2000). Michael Gorman, for example, insists that

> There is a dearth of research in US LIS schools that is dedicated to the real needs of real libraries. This is the result both of the divorce between information science oriented faculty and practicing librarians and of the fact that LIS schools in the US tend to be part of large universities that value (and reward) pure research over applied research. This has led to a gap in the library journal literature between arid and inaccessible reports of pure research and naïve "how we did it good" reports. (Gorman, 2003)

Indeed, fostered by the factors identified earlier, the education establishment for this field is in the midst of a period of change which has led to major repositioning and focus, major curricular changes, adoption of new approaches, technologies and knowledge, and identification of new constituencies and an extensive infusion of new resources in some schools. The information technology revolution provided the necessary impetus for change and programs with faculty attuned to these external factors and engaged in research are more likely to be positioned for this essential change.

IV. Documenting LIS Education in the Midst of Change

W.K. Kellogg, the breakfast cereal industry pioneer, established the W.K. Kellogg Foundation in 1930. Since its beginning Kellogg Foundation has continuously focused on building the capacity of individuals, communities, and institutions to solve their own problems. It seeks "to help people help themselves through the practical application of knowledge and resources to improve their quality of life and that of future generations." (Kellogg Foundation)

Seeing the Internet revolution in its infancy and fearing a crisis in the delivery of information by libraries and in education for librarians that would result in these institutions falling hopelessly behind, the Kellogg Foundation developed a program initiative, Human Resources for Information Systems

Management (HRISM), designed to insure that information professionals would be able to increase access to "knowledge and resources" with the aim of improving the quality of life for people. The concern at the time was that, without intervention, the Internet revolution might very well render libraries and librarians irrelevant (Bishoff, 1999).

During the tumultuous period of the 1990s, the W.K. Kellogg Foundation, seeking to influence change in LIS education, funded several experiments in change among a group of LIS programs including the University of Michigan, Florida State University, Drexel University, and the University of Illinois. The University of Illinois Graduate School of Library and Information Science (GSLIS) grant focused on the revision of the core curriculum courses (shortly after developing an extensive distance education program). Florida State University's LIS program (which has since become the School of Information Studies), focused on the development of an undergraduate degree in information technology and network management. Drexel University's College of Information Science and Technology focused on the use of information technology in curriculum delivery. The University of Michigan changes will be profiled under Trend 1, below.

At the end of the HRISM experiments, Kellogg, in addition, funded the most extensive examination of LIS education in nearly a century—KALIPER, the Kellogg–ALISE Information Professions and Education Renewal Project. The findings of KALIPER and additional developments in the past 4 years that impact the education of those who seek to become librarians are discussed below.

V. The KALIPER Project

The 1997 conference of the Association for Library and Information Science Education (ALISE), entitled "Reinventing the Information Profession," featured interdisciplinary speakers, highlighting some of the results of the Kellogg educational experiments mentioned above. (Durrance and Pettigrew, 1999, p. 287). This conference also received funding from the Kellogg Foundation and Kellogg leadership was present. Following the conference, a group of ALISE leaders approached the Kellogg Foundation for additional funds to look broadly at educational changes being made at schools of library and information science. KALIPER, a research project that was undertaken between 1998 and 2000 sought to

- determine the nature and extent of major curricular change in LIS education across North America, and
- help move curriculum reform toward achievement of critical mass in the field. (KALIPER Advisory Committee, 2000)

The study itself was a comprehensive examination of LIS curriculum conducted by a team of 20 scholars. The original aim of KALIPER was to review approximately one quarter of North American LIS programs. However, the interest in educational change was so great that KALIPER scholars, instead, examined nearly half the North American LIS programs looking for evidence of change in LIS education.

The KALIPER Project included: guidance by a Blue Ribbon Committee of field leaders; competitive selection of scholar researchers; an iterative study design incorporating multiple data collection methods starting with a dean survey to which 84% of all programs responded; case studies of the four Kellogg-funded programs; examination of a broad group of LIS/IS programs and comparison across schools; and analysis of statistics provided by KALIPER schools. It used a qualitative approach to analysis of programs using a variety of data collection methods including examination of program web sites, self-study reports, ALISE statistical reports, syllabi and readings for core courses, and selected interviews. The project, in addition, fostered exchange of data by scholars (Durrance and Pettigrew, 1999; Pettigrew and Durrance, 2000; Cox *et al.*, 2001).

The KALIPER Report, the most important study of LIS education since the Williamson Report (1923), was issued in 2000 and its Executive Summary was widely distributed both in paper copy and on the Internet. Reports on the project were given at major associations such as ALISE, American Society for Information Science and Technology (ASIST), the International Federation of Library Associations (IFLA), and various national and international conferences. A number of articles presented the KALIPER findings; two appeared in the *Bowker Annual Library and Book Trade Almanac* (Durrance and Pettigrew, 1999; Pettigrew and Durrance, 2000). ALISE devoted an entire issue of the *Journal of Library and Information Science Education* (Summer, 2001), and articles on the study have appeared in major library journals and on the Internet (Pettigrew and Durrance, 2001). Unlike the Williamson Report which found early library education in disarray, KALIPER found active movements toward change in the education of information professionals for libraries and other information environments.

Now, less than 4 years since the completion of KALIPER, it is clear that the trends first presented in 2000 have become even more pronounced. At the same time there has been growth in the formation of information programs arising outside of LIS. This section of the paper will examine the changes in LIS programs identified by KALIPER and highlight additional changes in the past 3 or 4 years, while the final section of the paper will examine and discuss rivals for the domain that has been claimed by LIS for many decades.

VI. KALIPER-Identified Trends and Their Current Manifestation

A. TREND #1: In Addition to Libraries as Institutions and Library-Specific Operations, Library and Information Science Curricula Are Addressing Broad-Based Information Environments and Information Problems

KALIPER scholars learned that by the end of the 20th century, LIS education had begun the change from a library-focused Ptolemaic model to an information-focused Copernican paradigm. Even before the Internet, a group of LIS programs reflected a rapid adoption of an information focus. KALIPER scholars found that faculty were very much aware that information professionals need to develop a "big picture" view of the information world. Curricula included courses framed toward broad information environments. KALIPER scholars found that schools were marketing both to a diverse student body and a diverse set of employers without, in the process, eliminating libraries as job targets for their graduates.

KALIPER teams found that LIS schools proclaim their domain as covering cognitive and social aspects of how information and information systems are created, organized, managed, priced, disseminated, filtered, routed, retrieved, accessed, used, and evaluated. How people get and use information has an increasingly prominent role in the curriculum with courses on user-centered design of information retrieval systems, information search strategy, and information-seeking behavior. LIS programs are incorporating approaches that deal with a variety of new problems into the curriculum including those associated with traditional content with an eye to increasing access to users. Courses look broadly at information access questions, redefining collections to better incorporate the virtual, and recognizing the blurring of institutional boundaries.

B. A Post-KALIPER Look at Trend 1: A Move to Information Programs

In the intervening 4 years since the KALIPER report, various LIS programs have moved even more rapidly toward more effectively addressing broad-based information environments and information problems in their programs. A quick look back at previous decades puts this move into perspective.

Thirty years ago, light years in the rapidly changing cyber world, Syracuse University, under the leadership of Robert Taylor, became the first "information school," by not only including information in the name of the program, but by also becoming the first school to drop the designation "library" entirely from its name, well over two decades before the Internet crisis of the 1990s. Indeed, Syracuse's website and other promotional materials proclaim it to be "The ORIGINAL Information School." Other library school programs began to incorporate "information" into their names 25 years ago, achieving a near total shift from "library"-only focused names to "library and information" (or "information and library") designations by the end of the 1980s.

It was not until the past decade, however, that programs began to drop "library" from their names in large numbers. By the mid-1990s only a handful of programs had eliminated library from their names. These actions were at least partially responsible for the American Library Association's First Congress on Professional Education (1999) (Kniffel, 1999). The move to eliminate the "L" word has continued at an accelerated pace. By early 2004 nearly one-third of accredited LIS programs (16 of 52) have chosen to remove "library" from their official names—and the trend continues. These actions have caused much concern among librarians that in removing "library" from their names, programs had, in the process, eliminated the institution from the curriculum. Sutton suggests that the "focus of study has shifted from the institution to the processes that underlie information creation, storage, transfer and use" and, in the process, education has been strengthened (Sutton, 1995).

While decades in coming, this trend signifies that educational programs have taken a turn in the road that has, indeed, resulted in a distancing from its origins—the need to educate professionals for libraries. At the same time this trend strengthens the education of information professionals, whose numbers include librarians, by focusing more broadly on information problems and environments (such as the Internet). In less than 10 years a group of "schools of information" or "I-schools" have emerged from schools formerly called schools of library and information science. These include, in addition to Syracuse, schools at the University of Michigan, the University of Washington, Florida State University, the University of California, Berkeley, the University of California, Los Angeles, and the University of Texas, Austin. The new programs are experimental, and while sharing a number of similarities no two are alike. Only time and the evolutionary process will reveal the most successful experiments. One such experiment at the University of Michigan is described below.

The University of Michigan School of Information and Library Studies (SILS) had a highly regarded MLIS program in the early 1990s. Incoming dean, Daniel E. Atkins, led the faculty in the submission of a proposal to Kellogg (Educating Human Resources, 1996). The grant proposal identified the strong need to create radical changes in education for information professionals; it identified both strengths and weaknesses in the intellectual constructs of several disparate fields, including LIS, computer science-engineering, and management information systems. It then proposed a model for information education that would mobilize the strengths of multiple disciplines.

Kellogg, understanding that LIS programs, including Michigan, had been under-funded, provided a major infusion of funds. The University of Michigan, as well, made substantial financial investments in the School of Information between 1992 and present. This enabled the School to assemble a core of interdisciplinary faculty and begin the difficult process of developing a curricular-scaffolding nexus that spanned several disciplines before it was necessary to show large enrollment increases. In this way, the School was able to create the instructional vision first and then attract students to it, rather than the more common, but less effective practice, of trying to co-evolve a pedagogical vision and rapid student growth. These funds also enabled the development of a substantial information technology infrastructure as well as the creation of infrastructure in the areas of student services, instructional computing, and teaching innovations that would have been impossible otherwise.

The School of Information, re-chartered by the University of Michigan Regents in 1996, was conceived as a non-departmentalized enterprise focused on an integrated learning model with multidisciplinary foundations, symbolized by Borromean rings focusing on the three basic domains of interest to SI, and their interlocked nature. In the rendition below they are flattened and resemble a Venn diagram, emphasizing both their separate parts and their binary and tertiary interactions (Fig. 1).

The multi-disciplinary SI faculty are united by several shared beliefs. One is belief in an increasing need for *information professionals*, skilled

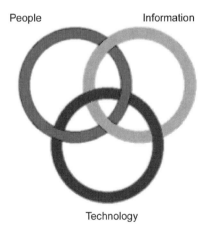

Fig. 1 Borromean rings representing people–information–technology.

practitioners who understand the complexities of the information realm who can help individuals and organizations deal with those complexities in reaching their objectives. This is not surprising, given that this professional bias was embodied in the SILS faculty and in the fields from which new faculty came: management, public policy, engineering, computer science, etc.

SI's vision is that of a heterogeneous and multi-disciplinary faculty tied together by shared interests and a common commitment to professional education and research focused on information. The result has been a rich portfolio of research and instructional capacities that spans disciplines and breaks new ground. The integrated vision provides the master narrative of the School, but it is embodied through practical specializations in Library and Information Science (LIS), Human–Computer Interaction (HCI), Archives and Records Management (ARM), and Information Economics, Management and Policy (IEMP). These serve as formal specializations within the integrated Masters of Science in Information (MSI) degree program, and they help organize activity within the School and establish identifiable linkages to professional communities outside the School. They have become strong mobilizing forces without eroding or threatening an integrated vision.

The framework of specializations at information schools like the UM School of Information bears watching. For example, the 25 or so other ARM programs in North America are embedded as sub-specializations in LIS programs or history departments. SI is the first school to offer graduate education in ARM that is on par with other specializations. The larger ARM community is watching this development with interest. The younger field of HCI, in contrast, has wandered from one possible institutional home to another, finding difficulty in establishing legitimacy in programs such as engineering, computer science, management, or psychology. The HCI community has watched the evolution of HCI in SI, and a number of strong HCI groups have begun to emerge in a group of newly formed information schools evolving along the lines of the SI model.

Changing the pedagogical thrust of education in the information professions requires simultaneous change in the larger institutional realm of the professions. This lesson was learned quickly upon the launch of the MSI program in 1996. The American Library Association's Committee on Accreditation had scheduled the re-accreditation review for the MILS for 1997. The relatively simple solution of seeking accreditation only for the library specialization within the MSI degree was antithetical to the whole idea behind the new degree. The School worked with ALA on an innovative scheme to accredit the entire MSI degree, thereby broadening the scope of what might constitute professional training in librarianship and enabling students with much broader training to take positions that required a graduate degree from an ALA-accredited program. Five-year accreditation for the entire MSI program was granted by ALA in 1998. This action sparked controversy among traditionalists within the profession. In the end, however, the innovation prevailed: in 2002–2003 the ALA reaccredited the MSI for a full 7 years, specifically noting SI's leadership in broadening the field of library education.[1]

C. A Post-KALIPER Look at Trend #2: While LIS Curricula Incorporate Perspectives from Other Disciplines, a Distinct Core Has Taken Shape that Is Predominantly User-Centered

Trend 2 addresses two important and related areas; it encompasses both increased user-centeredness and increased interdisciplinarity (often bringing different disciplinary views of the user). The missions of most LIS programs as well as the emerging Information Schools show these academic programs to be user-centric. The University of Michigan's School of Information's core mission, for example, is based on an integrated approach to the study, design, and management of information systems, in particular "bringing people, information, and technology together in more valuable ways" [Mission Statement].

There has been an infusion of multidisciplinary perspectives into LIS curricula as LIS faculty have broadened their focus beyond libraries, as faculty from multiple disciplines are hired, and as faculty conduct research with colleagues who have degrees from other fields. These perspectives emerge as well when schools offer joint programs and courses or team-teach with faculty from other departments. Faculty in increasing numbers of LIS programs are growing increasingly multidisciplinary with new hires and through additional joint appointments.

Information-focused programs focus on individuals, groups or societies. While employing a user-centered perspective has been a hallmark of some schools' curricula for a long time, there is little doubt that "user-centeredness" has infused or pervaded most of the research and teaching in LIS. Extensive conceptual and empirical research focusing on information seeking and use, as well as user interaction with information systems, has made strong contributions to the knowledge base and, as a result, to curriculum (Kuhlthau, 2004; Pettigrew et al., 2001). An increasing number of core courses or course components address information seeking. In revisions of core courses, the incorporation of instruction in information seeking could be seen in varying degrees of granularity ranging from the cognitive issues of personal information seeking and use to the broad-based role of information in practice and discourse communities. For example, increasingly schools have added

---

[1] This brief description of the University of Michigan School of Information has been adapted from the strategic assessment document prepared by the School in December 2003.

faculty whose interests focus on HCI which focuses on designing, developing, and evaluating technologies that fit the capabilities of the user, the work to be done, and the surrounding work practices and organizational context.

D. A Post-KALIPER Look at Trends 1 and 2: Expanded Interdisciplinary Research Focusing Broadly on Information Problems and Environments; the Development of a Golden Age in Scholarship

While KALIPER addressed a variety of curricular and support questions, it did not directly address research. The information revolution provided the opportunity to broaden scholarship. Although this influence was seen and reported at the time of the KALIPER report, it has expanded, in part because of an increasing number of new hires from fields that also focus on aspects of the changing information landscape and in part because the changed information environment has provided LIS researchers with the opportunity to apply their research approaches to broader information problems and environments.

Moreover, the move from a library-centered paradigm to an information-centered paradigm and the increased interdisciplinarity both of the new information schools and LIS programs has resulted in an increased ability to identify frameworks that explain the types of research conducted by LIS faculty. Figure 2, "Broad Groupings of LIS/IS Research" is based on an examination of program websites that feature faculty research as well as a general examination of LIS/IS research in such texts as Rubin's (2000) *Foundations of Library and Information Science*. Figure 2 groups this research into five broad categories—information technologies, content, information systems, human information behavior, and cross-cutting categories. It reveals the breadth of contemporary research interests across a wide range of information environments and information problems. Figure 2 shows that LIS researchers look broadly at problems associated with increasing access to information; it does *not* suggest that researchers ignore libraries; rather it suggests the variety of topics that inform LIS education—and thus librarians.

UCLA faculty member Marcia Bates recently noted the broad applicability of LIS/IS research. Writing both for LIS audiences and more broadly, Bates charges that while building the Internet,

> hundreds of millions of dollars have been invested to re-invent the wheel—often badly. Everybody understands and takes for granted that there is an expertise needed for the application and use of technology. Unfortunately, many Web entrepreneurs fail to recognize that there is a parallel expertise needed about information—collecting it, organizing it, embedding it successfully in information systems, presenting it intelligently in interfaces, and providing search capabilities that effectively exploit the statistical characteristics of information and human information seeking behavior. (1999)

Information Technologies	Information/ Knowledge (Content)	Information Systems	Human Information Behavior	Cross-Cutting Areas
Technology capabilities and limitations Historical aspects (including various information technology innovations) Issues; legal questions Impacts of IT Identifying and selecting information technologies Human factors in technology Specific information technologies such as the internet and web technologies Cyber-infrastructure	Defining the nature of information and its value Life-cycle of information Publishing (including electronic) Physical and virtual collections Economics of information Costing and pricing of information and information services Value-added functions Bibliometrics; webmetrics	Information storage and retrieval Computerized information systems User-centered design of information systems Approaches to organization of knowledge/ information Increasing system capabilities Search retrieval models Database and file structure Computer–human interfaces Expert systems & intelligent agents Studies of use of the system or information resources	Information needs information seeking and search processes Characteristics of information users Information use and uses Human information interaction Information literacy Impacts (outcomes) of information use Effects of information on decision-making Communication and professional practice designed to increase access to information (including service development)	Information Environments Historical aspects Management approaches and concerns Evaluation approaches and issues Information policy Methods

Fig. 2 Broad groupings of LIS/IS research. This figure is influenced both by examination of individual research profiles of LIS faculty on School websites and Chapter 2 of Richard Rubin. *Foundations of Library and Information Science,* NY: Neal-Schuman, 2000 (especially pp. 23–53).

Bates points out that while librarians have "created multi-million-item online public access library catalogs, when online access was a brand-new concept, and had developed a tremendous amount of expertise about how to handle large, messy databases of textual information…it has been almost an article of faith in the Internet culture that librarians have nothing to contribute to this new age" (Bates, 1999).

Bates posits that much of the expertise of information science is invisible and, as a result, people outside the field, including researchers from other domains, fail to recognize the contributions of this meta-field whose domain "is the universe of recorded information that is selected and retained for later access" as such "cuts across, or is orthogonal to, the conventional academic disciplines. (1999, pp. 1044, 1046)" Bates argues that because much "information work" is unseen, it is thus undervalued in the information environment characterized by the Internet. She adds that as "the society at large is discovering information and the problems of information description and organization" but failing to recognize the expertise required to describe and organize knowledge (1999).

E. An Emerging Golden Age of Broadly-Framed Monographs

The field's more broadly-focused research as well as the increased ability of its researchers to articulate the domain of LIS, has resulted in what could be considered a golden age of monographs that serve to distill and integrate the knowledge base that has been amassed by LIS/IS faculty. The monographs identified below, all published within the past 4 years, are indicators of a golden age of scholarship, precipitated, at least in part by the changes in the information landscape that have helped scholars more effectively articulate the core knowledge of the field.

Christine L. Borgman (2000), Professor and Presidential Chair at the UCLA Graduate School of Education and Information Studies, is the author of *From Gutenberg to the Global Information Infrastructure: Access to Information in the Networked World*. This book brilliantly examines the emerging global information infrastructure (GII). Borgman presents a big picture view of changes associated with digital libraries and the Internet. She poses the construct of a "global digital library" as a framework for thinking about "access to information in an internationally distributed computer network." In the closing paragraphs of the book Borgman states,

> Research on digital libraries and on access to information has moved from computer and information science into the physical and life sciences, the social sciences, and the humanities. Concurrently, research questions have expanded from technical concerns for information retrieval and content representation into social aspects of digital libraries and across all phases of the information life cycle. Scholars in the disciplines are working with computer and information scientists to construct and study digital libraries tailored to their information needs and practices. Researchers are partnering with information professionals such as librarians, archivists, curators, and records managers to address pragmatic technical issues, management questions, and preservation and policy concerns.

Geoffrey C. Bowker and Susan Leigh Star (2000) who were for several years on the faculty at the University of Illinois GSLIS, have written an excellent monograph on the meaning and uses of classification. A reviewer of the book, *Sorting Things Out: Classification and its Consequences* asks:

> Is this book sociology, anthropology, or taxonomy? *Sorting Things Out*, by communications theorists Geoffrey C. Bowker and Susan Leigh Star, covers a lot of conceptual ground in its effort to sort out exactly how and why we classify and categorize the things and concepts we encounter day to day. But the analysis doesn't stop there; the authors go on to explore what happens to our thinking as a result of our classifications. With great insight and precise academic language, they pick apart our information systems and language structures that lie deeper than the everyday categories we use. The authors focus first on the International Classification of Diseases (ICD), a widely used scheme used by health professionals worldwide, but also look at other health information systems, racial classifications used by South Africa during apartheid, and more.   (Lightner, 2000)

Two LIS faculty members, Ann Peterson Bishop of the University of Illinois, GSLIS and Nancy Van House of the University of California, Berkeley, School of Information Management and Systems, have collaborated with a Geographic Information System (GIS) researcher to edit an excellent monograph on approaches to evaluating digital libraries, *Digital Library Use: Social Practices in Design and Evaluation* (Bishop et al., 2003). T.D. Wilson, commenting on this landmark book, notes,

> It will be readily apparent that, at the present stage of development of digital libraries, a sociotechnical systems perspective ought to be productive of inter-disciplinary approaches to problems. And so it appears to be the case in this volume. The authors of the papers included here are from a variety of different disciplines… [including] names recognized in various aspects of computer science, communication studies, human/computer interaction studies, anthropology, librarianship and information science. In fact, when the digital library phenomenon is reviewed fifty years from now, it may be recognized that its key contribution will have been to lift library research out of its self-defined 'ghetto' and into the wider world of scholarship.   Wilson (2004)

The reviewer concludes that, "this volume ought to be made essential reading for any librarian, any library researcher and any academic in the field."

The long-awaited 2nd edition of Carol Collier Kuhlthau's groundbreaking book, *Seeking Meaning: A Process Approach to Library and Information Services*, has just been issued (2004). This work in its earlier edition has made very strong contributions to shaping research in information behavior and bringing the field to maturity through her theoretical contributions. One of the major contributions of her empirical research is documentation of change in the holistic experience of people in the process of information seeking, incorporating the physical, cognitive and affective dimensions from the perspective of the user. At the time that it first appeared in the early 1980s it was a radical departure from the source and point of access approach of much of the research in the field. Kuhlthau's work is as widely admired by practitioners as it is by researchers and theoreticians.

The appearance of University of Kentucky School of Library and Information Science faculty member Case's (2002) *Looking for Information: A Survey of Research on Information Seeking, Needs, and Behavior*, reveals the maturity of scholarship associated with research on information behavior. Case's excellent examination of scholarship in this area was named the winner of the 2003 ASIST Best Information Science Book Award. Case reviews hundreds of studies of information-seeking behavior and examines the body of research on information seeking, including basic research on human communication behavior as found in the literature of psychology, anthropology, sociology, and other disciplines. This book would not have been possible without the cumulative research of scores of researchers. Case's fine book can serve as an excellent introduction for practicing librarians to the growth of scholarship in this vital area.

The final book cited in this brief list of broadly framed monographs that are becoming LIS text books—and which should be required reading for all practicing librarians—is by Professor Richard L. Rubin, Director of the School of Library and Information Science at Kent State University— *Foundations of Library and Information Science*. NY: Neal-Schuman, 2000. Rubin's introduction to LIS seeks to bring together the full range of LIS research (2000). The quoted segments from a group of interviews, below, is taken from Amazon.com samples of the many reviews of this title.[2]

Journal of the American Society for Information Science
"Sound and thorough. Rubin's book will for now be the textbook for MLIS foundation courses"
Journal of Academic Librarianship
"[An] important addition to the library and information science (LIS) education literature"
Library Quarterly
"Rubin helps present and future librarians understand the need to respect the past but to prepare for the future"

This book, like Case's, would not have been possible without the growth of research and scholarship in LIS. It is evidence of the maturity of the field in this first decade of the 21st century.

The six titles included here are only a few of the rich array that have appeared within the past 4 years. They are reflections of an extraordinary period of LIS scholarship and knowledge growth; they place LIS properly as

---

[2] http://www.amazon.com/exec/obidos/tg/detail/-/1555704026/104-9704125-8276755?v = glance&vi = reviews, accessed April 24, 2004.

a key contributor to knowledge growth beyond what has been considered the domain of librarians. These titles show the value of examining phenomena from a broad vantage point.

The final set of trends identified by KALIPER, discussed below, are outgrowths of this broad view of the field.

F. Trend #3: LIS Programs Are Increasing the Investment and Infusion of Information Technology into Their Curricula

KALIPER noted that the increase in investment in information technology infrastructures and the infusion of information technology into the curricula should not be simply dismissed as a sign of the times. Something more meaningful is occurring that is having long-reaching effects. The intense focus on most anything digital is undoubtedly redefining LIS education as we add more core courses and electives to the curriculum, infuse existing courses with digital elements, and seek out more faculty who can teach in these areas. Information technology is attractive, it is fast becoming the glue of our daily existence, and market forces and funders of education and research are willing to support IT development and use. For these same reasons, the parent institutions want programs that lead in teaching and research on the electronic frontier.

G. A Post-KALIPER Look at Trend 3: Leadership in Cyber-Infrastructure Growth

Information technologies continue to explode requiring LIS programs to continue IT development and to hire faculty capable of incorporating both knowledge and skills into the curriculum. Some schools such as those who participated in the federally funded digital library initiatives are conducting research for cyber-infrastructure development, the comprehensive, advanced infrastructure based on information and communication technology, including the Global Information Infrastructure and preparation for the next generation of information technologies. Faculty in LIS and information schools continue to make strong contributions to the knowledge base in this area (Borgman, 2000). Professor Dan Atkins at the University of Michigan School of Information recently chaired the National Science Foundation's Blue Ribbon Panel on Cyber-Infrastructure that produced a major set of recommendations that are expected to have wide-ranging impact on the cyber-infrastructure of the United States (Revolutionizing Science, 2003).

H. Trend #4: LIS Schools and Programs Are Experimenting with the Structure of Specialization within the Curriculum

Schools involved with KALIPER indicated that they were "rethinking specializations" including offering more generic curricula, adding new specializations such as medical informatics, or developing joint degrees with other schools.

As part of preparing students for specialization, some schools impose program entry and/or exit requirements, such as work experience in industry, or require their students to complete practical engagements or compile graduation portfolios that describe their field experiences during their programs. Other exit requirements include successfully completing internships or other practical engagement activities.

I. A Post-KALIPER Look at Trend 4: Experimenting with the Structure of Specialization

In a number of programs, with the exception of school library media specializations, there is less emphasis on type of library specializations. Promotional materials developed by LIS programs often indicate that students are provided a generalist education so that students will be prepared to work in a variety of environments. For example the University of Washington School of Information indicates that the "Master of Library and Information Science program is a 63-quarter-credit program which takes a generalist approach while still offering numerous opportunities for students to focus on particular areas of emphasis." (University of Washington)

A new set of specializations are beginning to replace the type of library specializations. Often these focus more on the needs of users. Emporia School of Library and Information Management, for example, indicates:

> A professional program must be designed to meet the changing needs of the profession and of society in general. SLIM is constantly reviewing and, when necessary, restructuring its curriculum to meet these changing needs. The curriculum focuses on the varied needs of diverse users and on interpersonal interactions and communication as essential elements in the design and implementation of information services. (Emporia)

Similarly the University of North Carolina—Chapel Hill Program Presentation indicates that possible specializations at the School of Information and Library Science include: "human information behavior, database and information retrieval systems, networking and Internet technologies, and management of information systems." (University of North Carolina) Rather than limit the student to one environment, these new specializations provide students with knowledge and skills they can use in a variety of settings.

J. Trend #5: LIS Schools and Programs Are Offering Instruction in Different Formats to Provide Students with More Flexibility

Flexibility in the curricula is perhaps nowhere as evident as in instructional formats. Today's students have more choice than ever regarding course length, day and time of course offering, and on or off-campus meetings. Traditionally, distance education courses were offered in a different physical location; within the past few years there are an increasing number of off-campus courses offered via some form of telecommunication and/or via the Internet.

K. A Post-KALIPER Look at Trend 5: Distance Education

Ten years ago only 10 North American schools offered courses using the distance education option. Now 36 (well more than half of all accredited LIS programs) provide technology assisted distance-education degrees (Daniel and Saye, 2001). The ability to use information technologies has revolutionized the way that LIS education is delivered. The Syracuse University School of Information Studies website indicates that the School "has been offering master's degrees through distance learning since 1993" noting that "online courses are of the same academic rigor as on-campus courses. Members of the School of Information Studies faculty teach in both the online and on-campus formats." (Syracuse) The number of programs delivering the LIS degree distance education is so extensive that the phenomenon itself has spawned a body of research. (Illinois)

L. TREND #6: LIS Schools and Programs Are Expanding Their Curricula by Offering Related Degrees at the Undergraduate, Master's, and Doctoral Levels

KALIPER scholars documented the rapid expansion of undergraduate programs, noting that the rapid enrollment gains in a number of LIS programs was due to the growth in related degree programs identified by KALIPER, particularly at the undergraduate level. The KALIPER study also noted the growth of additional master's and doctoral programs. Pettigrew and Durrance, in a summary of the project, commented:

> In short, schools are expanding in many directions. New continuing education programs, workshops, and other alternative programs have enabled schools to tap into expanded markets and provide another potential source of revenue. Since its merger with the College of Education, Missouri has implemented a certificate program in new media at the undergraduate and graduate levels. South Carolina offers two post-master's programs—certificate of graduate study in library and information science, and specialist in library and information science—while Syracuse offers a summer college for high school students in information management and technology. Syracuse also increased its number of graduate

certificates to include telecommunications management and software project management with a possible addition of interactive multimedia. (Pettigrew and Durrance, 2000, 2001)

M. A Post-KALIPER Look at Trend #6: LIS Schools and Programs Are Expanding Their Curricula by Offering Related Degrees at the Undergraduate, Master's, and Doctoral Levels

Since the KALIPER report was issued, a number of schools have developed or are developing innovative undergraduate programs (majors and/or minors). For example, undergraduate degrees are offered in such areas as: Information Technology; Information Science; Information Systems; or Information Technology and Informatics. Those receiving undergraduate degrees comprise more than a third of the graduates of Drexel, Pittsburgh, and Syracuse and half of the graduates at Florida State University and the University of Wisconsin, Milwaukee (ALISE, 2002).

Some programs have differentiated master's degrees. For example,

- University of North Carolina offers a Masters of Science (MS) in Library Science and an MS in Information Science.
- Rutgers, likewise, has two degrees, the Master of Library and Information Science; and the Master of Communication and Information Studies.
- Syracuse, in addition to its MS in LIS has an MS with School Media Certification; an MS in Information Management, an MS Federal Government Specialization in Information Management (in Washington, DC), and an MS Telecommunications and Network Management. Syracuse's Masters in Information Science accounts for 31% of its graduates. (ALISE Statistics, 2002)

In sum, the changes identified in North American LIS programs by KALIPER scholars have continued and accelerated, thus shaping LIS education for the new digital era. The most noticeable changes have been in increased interdisciplinarity, the move toward curricular developments and research that focus broadly on information problems and environments, and a recent move toward the development of newly emerging information schools, a phenomenon to be discussed below.

Importantly, KALIPER has thus influenced the ways LIS is framed. The findings, themselves, have had an impact beyond bringing together the data on change. They have been: incorporated into articles that discuss curricular change, used in curricular revision in various LIS programs, discussed and debated by librarians, and used as the basis for new "KALIPER" studies in other countries.

Programs seeking to educate information professionals, including librarians, for the 21st century are stronger than ever. Students better understand the needs that people have for information and how to more effectively assist them in getting the information they need, they gain skills in using information technologies, and they have a broad understanding of information systems. These changes not only have prepared LIS education for the digital age, they have also moved it toward a convergence of various disciplines, each making some claim on control of the domain.

VII. Information Education: Competition or Convergence?

Modern education for librarianship at the end of the 19th century arose in response to what was then a major crisis in staffing brought on by the generosity of Andrew Carnegie and his Carnegie Corporation that resulted in over 2000 public library buildings (Van Slyck, 1995). The Carnegie Corporation, recognizing the unintended outcomes of its largesse, commissioned a study of the approaches to educating the librarians for these newly funded libraries. The study, called by the name of its developer—the Williamson Report—uncovered a crisis of education (Williamson, 1923). This study showed that the educational apparatus in place in the early 1920s was grossly inadequate; many librarians trained on the job, most educators were ill-prepared, courses were too rudimentary, there was no consistency in training, the field lacked textbooks, there were no standards to assure quality, and education itself was grossly under-funded (Williamson, 1923). To protect its investment, Carnegie fostered the institutionalization of library education in universities and moved toward an accreditation process that would improve the dismal quality found by Williamson. It is important to note, however, that both the crisis and its solution were library centered: the solution to the crisis brought on by too many poorly trained librarians was to create educational programs designed to produce better educated librarians. This narrow construct, focusing education on a single institution, did, indeed, solve the immediate problem. However, over time—as other disciplines recognized the value of organizing the world's *digital information*, this institution-specific solution contributed to sowing the seeds for the current crisis in LIS education described by Van House and Sutton (1996) as the "Panda Syndrome."

The KALIPER Report clearly demonstrated that educators fully understood the value of providing librarians the skills they needed to organize and retrieve information on the Internet (as well as in other formats). In the process of change, LIS education has taken great strides to assure that its graduates are capable of anticipating and responding to the needs of the digital age; these changes have brought the convergence with other disciplines that focus on digital formats.

Today conditions are ripe either for the field's most serious crisis or for an unparalleled growth opportunity. The changed information landscape has created an uneasy playground for a disparate group of players that results in both a threat and promise of a renaissance in librarianship as one of an emerging group of information professions. Various competing players, each breaking out of formerly narrow constructs, have laid claim to the same domain–information. The Internet crisis has resulted in a new set of information life cycle problems that need to be solved (Hodge, 2000); various solutions are being offered by professionals educated in different disciplines.

Van House and Sutton (1996) warned of converging and competing interests. These interests, of course, include education programs that stand ready to educate some segment of tomorrow's information professions. For example, with the rapid rise of the personal computers in the 1980s and the urgent need to improve computer interfaces for non-computer scientists, the sub-field of computer science now known as HCI emerged. Its primary professional organization, the Computer–Human Interaction (CHI) developed in the early 1980s as a Special Interest Group of the Association for Computing Machinery (ACM). Well-developed HCI programs are in universities such as the MIT Media Lab, Carnegie Mellon's HCI Institute, Georgia Tech, Virginia Tech, and the University of Maryland. Scores of programs are emerging from computer science that focus on preparing students broadly for careers in information technology. The School of Information and Computer Science was formed in 1968 at the fledgling University of California, Irvine as one such program. These programs are becoming increasingly interdisciplinary with the realization that the knowledge from a single discipline is inadequate either to conduct research or to develop relevant curricula. Other academic programs designed to educate information professionals include various informatics programs, most commonly medical informatics in Medical Schools and information management programs in Business Schools.

Major digital library initiatives funded by the US government brought together researchers from various disciplines and fostered interactions among computer scientists, LIS researchers, economists, and others such as experts in GIS. These interactions have provided a vehicle for examining disciplinary differences and, as a result, researchers across several fields have come to a better understanding of their differences and have begun to develop "a view [of digital libraries] that encompasses the social, behavioral, and economic contexts in which digital libraries are used." (Borgman, 2000, p. 240) Rival claims of jurisdiction over the domains associated with the new information world coupled with pressures from constituents in libraries could nudge LIS education into its former narrow library focus. However, intellectual convergences, such as those that resulted from the federally funded digital library initiatives, can be seen among a diverse group of researchers, many of them in the newly formed interdisciplinary "information" schools.

The most recent convergence examples that have emerged from the new information landscape are cross-disciplinary experiments such as the one at the University of Michigan School of Information. Such new schools hire faculty from multiple disciplines. The new information schools have hired PhDs from computer science, HCI, information economics, cognitive psychology, LIS, and other fields. For example, University of California at Irvine has sent PhDs to several of the new information schools and a dean to the University of Michigan. Penn State, a newly formed information school with no history of LIS, has hired faculty with LIS degrees. Since these new programs have now been in existence for between 5 and 10 years, they have begun to graduate PhDs who are taking positions in LIS programs, computer science departments, or the emerging information schools.

Information schools that bring together researchers with exceedingly different intellectual fields with disparate cultures, values, methods, traditions, and approaches to knowledge development into a single faculty are faced with culture shock. Newly interdisciplinary programs need to take steps to overcome differences and build on the strengths of the various disciplines represented by their faculty. Academic integration at this stage of development is a challenge and has been established as a primary activity for such programs as the University of Michigan School of Information. The aim of academic integration is to move beyond the natural distrust that academics have of those educated in vastly different traditions.

Two very important convergence activities underway are discussed below. If successful, these efforts may help create a convergence that could result in the development of a new, interdisciplinary field that encompasses the strengths of multiple fields. The first, within LIS, brings together the new information schools with the aim of creating support systems that will make it easy for these new information-focused programs to discuss their common missions and problems. This initiative began with several meetings of deans from the University of Pittsburgh School of Information Sciences, Syracuse University's School of Information Studies, the Drexel University College of Information Science and Technology, the University of Washington Information School, and the University of Michigan School of Information. As the number of information schools expands, additional deans have begun to attend these meetings. The information school deans have held meetings with deans of all LIS programs since 2003 at the ASIST and at the ALISE. These meetings have focused on broad issues associated with the question of convergence made possible by the common interests of many computer science, engineering, LIS and management information systems programs, the emerging information school movement, and its impact of programs of LIS.

The second convergence initiative has convened a community of deans of information technology (IT) programs that have emerged from computer science departments and the newer, more broadly focused, interdisciplinary information schools. This group has been meeting under the auspices of the major US association focused on research in information and computing, the Computing Research Association (CRA) (www.cra.org) through a division known as the IT Deans Group (CRA-IT Deans Group). CRA started as a membership association of computer science departments, but has expanded to include other kinds of programs. The CRA meetings of the IT Deans Group reveal the struggles faced by the IT programs which have emerged from computer science because their research separate from computer science which, these researchers believe, has developed approaches and knowledge which "often do not live comfortably in departments of computer science." (Finkelstein and Hafner, 2002, p. 4) Thus, a group of IT researchers recognize the need for an emerging IT discipline, which would encompass topics such as these below:

- The study of information: how it is acquired, organized, communicated, managed and used by people and organizations, and how IT changes those processes, sometimes in fundamental ways.
- The study of IT applications per se, including application taxonomies based on technical requirements, functional characteristics, information models, and domain or context of use (e.g., business, government, education, health care, publishing, the military, law enforcement, media and entertainment, science and engineering).
- Techniques and tools for managing the design, development and deployment of large complex IT systems.
- The study of how IT affects human behavior and quality of life.
- The study of how IT affects social and political institutions, and how those institutions in turn affect the development and use of IT. (Finkelstein and Hafner, 2002, p. 4)

While some of these goals are dissimilar from those of LIS, others resonate with the goals of LIS research—and can be seen either as intellectual competition or convergence. While these researchers, however, do not appear to recognize the value of information, itself, nonetheless it is not difficult to see a further convergence that incorporates the research strengths of LIS such as those identified by Bates above.

The leadership of the CRA-IT Deans in bringing together researchers from different disciplines makes it possible for those with divergent perspectives and knowledge to communicate and, possibly, to collaborate. This initiative, which began only 4 years ago, has only begun the work of identifying problem areas and points of convergence

(CRA-IT Deans). At present about 40 programs, including a group of leading LIS/IS schools, are participating in the CRA-IT Deans meetings. These meetings focus on examining the implications of the convergence of various disciplines into a common domain and building a conceptual picture of the intellectual territory covered by all the research and instruction programs. A major conference to examine the intersection of interests represented by a variety of academic programs is planned for summer 2004. (CRA-IT Deans)

Convergences across various academic programs toward an information domain is depicted in Fig. 3: "Moving Toward Intellectual Convergence." The bold arrows show stronger organizational convergence which is seen in participation in interdisciplinary groups such as CRA, while the dotted arrows represent programs that have not moved as clearly toward collaboration with other disciplines.

Schools of information, whether their origin is from LIS or computer science are now seen as representative of an intriguing and powerful notion. An institutional transformation is under way in higher education focused on the study and design of information and information technologies. Key discriminating factors among these new programs are the degree to which they successfully pursue an interdisciplinary view of the problems, and their commitment to the goal of improving human welfare as a result of their efforts. Knowledge gains in LIS/IS leading to an integrative scholarship have assured that the next generation of librarians will be well prepared for the future and that LIS is positioned to become a strong player in the transformation of information education. Christine Borgman summarizes the recognition of a number of researchers in this period of convergence.

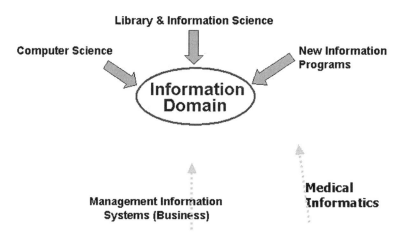

Fig. 3 Moving toward intellectual convergence.

Access to information is too important a problem to leave entirely to government officials, corporate policy makers, librarians, archivists, computer scientists, or lawyers. Rather it is a problem faced by people in all walks of life, at most stages of life, in all parts of the world.   (Borgman, 2000, p. 269)

References

Abbott, A. D. (1988). *The System of Professions: An Essay on the Division of Expert Labor*. University of Chicago Press, Chicago.
American Library Association. Congress on Professional Education (1999). *1st Congress on Professional Education: Focus on Education for the First Professional Degree* (June).
American Library Association. Congress on Professional Education (2000). *2nd Congress on Professional Education*, November 2000.
Association for Library and Information Science Education (ALISE). *ALISE Statistics*, 2002. Table II-3-a-1. Degrees and Certificates Awarded by ALA Schools, 2000–2001, pp. 138–140.
Bates, M. J. (1999). The invisible substrate of information science. *Journal of the American Society for Information Science* 50(12), 1043–1050.
Bishoff, Liz. (1999). *Leadership: Evaluation of the Kellogg Foundation HRISM Library Education Project* CLIR, Washington, DC, http://www.clir.org/activities/details/completed/hrism/hrism.html, accessed April 5, 2004.
Bishop, A. P., Van House, N. A., and Buttenfield, B. P. (2003). *Digital Library Use: Social Practice in Design and Evaluation*. MIT Press, Cambridge, MA.
Borgman, C. (2000). *From Gutenberg to the Global Information Infrastructure: Access to Information in the Networked World*. MIT Press, Cambridge, MA.
Bowker, G. C., and Star, S. L. (2000). *Sorting Things out: Classification and Practice*. MIT Press, Cambridge, MA.
Case, D. (2002). *Looking for Information: A Survey of Research on Information Seeking, Needs, and Behavior*. Academic Press, New York.
Computing Research Association. IT Deans Group, http://www.cra.org/Activities/itdeans/, accessed April 24, 2004.
Cox, R. J., Yakel, E., Wallace, D., Bastian, J. A., and Marshall, J. (2001). Archival Education in North American Library and Information Science Schools. *Library Quarterly* 71(2), 141–194.
Daniel, E., and Saye, J.D. Highlights of the 2001 ALISE Statistical Report with a five and ten year comparison of key data elements. http://www.ils.unc.edu/ALISE/2001/Highlights.htm.
Durrance, J. C., and Pettigrew, K. (1999). *KALIPER: A look at library and information science education at the turn of the new century*. (1999 Bowker Annual), R.R. Bowker, New York.
Educating Human Resources for the Information and Library Professions of the 21st Century. A Proposal to the W.K. Kellogg Foundation from the Faculty of The [University of Michigan] School of Information and Library Studies, 1996. http://www.si.umich.edu/cristaled/Kelloggproposal.html.
Emporia State University. School of Library and Information Management. Website http://slim.emporia.edu/degrees.htm#mls, accessed April 24, 2004.
Finkelstein, L., and Hafner, C. (2002). The Evolving Discipline(s) of IT (and their relation to Computer Science): A Framework for Discussion. Presented at the IT

Deans Group Meeting, Washington, DC, February 9–10, http://www.cra.org/Activities/itdeans/finkelstein.pdf, accessed April 5, 2004.

Gorman, M. (2003). Whither Library Education? Keynote Speech at the Joint EUCLID/ALISE Conference: Coping with Continual Change—Change Management in SLIS. Potsdam, Germany, July 31st, 2003, http://www.fh-potsdam.de/EUCLID/tmp/Gorman-keynote.doc, accessed April 5, 2004.

Hodge, G. M. (2000). Best practices for digital archiving: an information life cycle approach. *D-Lib Magazine* (January), accessed April 24, 2004.

KALIPER Advisory Committee (2000). *Educating Library and Information Science Professionals for a New Century: The KALIPER Report. Executive Summary.* Association for Library and Information Science Education (ALISE), Reston, VA.

Kellogg Foundation. Website. Frequently Asked Questions, http://www.wkkf.org/Programming/FAQ.aspx?CID = 271#97, accessed April 12, 2004.

Kniffel, L. (1999n). Practitioners, educators seek library's place in professional education. *American Libraries* **30**(6), 12–15.

Kuhlthau, C. (2004). *Seeking Meaning: A Process Approach to Library and Information Services*, 2nd ed., Libraries Unlimited, Westport, CT.

Lightner, R. (2000). *Sorting Things out: Classification and Practice*. MIT Press, Cambridge, Amazon.com; Review of Bowker, Geoffrey C. and Susan Leigh Star.

Lynch, M.J. (2002). Reaching 65: Lots of Librarians Will Be There Soon. *American Libraries* (March).

Pettigrew, K., and Durrance, J.C. (2000). KALIPER study identifies trends in library and information science education. In *2000 Bowker Annual*.

Pettigrew, K.E., and Durrance, J.C. (eds.) (2001). KALIPER: introduction and overview of results. *Journal of Education for Library and Information Science* **42**(3), 170–180. Entire issue devoted to KALIPER findings.

Pettigrew, K.E., Fidel, R., and Bruce, H. (2001). Conceptual frameworks in information behavior. In *Annual Review of Information Science & Technology*, (M. E. Williams, ed.). vol. 35, pp. 43–78, Medford, NJ: Information Today.

Revolutionizing Science and Engineering through Cyberinfrastructure: A Report from the U.S. National Science Foundation Blue Ribbon Panel on Cyberinfrastructure. Daniel E. Atkins, Chair, January 2003, http://www.communitytechnology.org/nsf_ci_report/.

Rubin, R. E. (2000). *Foundations of Library and Information Science*. Neal Schuman, New York.

Sutton, S. (1995). Keynote Presentation. California Academic and Research Libraries. 3rd Annual Conference, http://www.carl-acrl.org/Archives/ConferencesArchive/Conference95/sutton.text.html.

Sutton, S. A. (2001). Trends, trend projections, and crystal ball gazing. *Journal of Education for Library and Information Science* **42**(3), 241–247.

Syracuse University. School of Information Studies. Website http://istweb.syr.edu/academics/distance/index.asp, accessed April 12, 2004.

Tenopir, C. (2002). Educating Tomorrow's Information Professionals Today. *Information Today* (July–August), http://www.infotoday.com/searcher/jul02/tenopir.htm, accessed April 24, 2004.

University of Illinois. Graduate School of Library and Information Science. Website. [Distance Education] Research Articles, http://alexia.lis.uiuc.edu/gslis/degrees/leep_bib.html#research, accessed April 24, 2004.

University of North Carolina. School of Information and Library Science. Program Presentation, http://www.ils.unc.edu/daniel/COA/MSIS.html, accessed April 24, 2004.

University of Washington. Information School. Website http://www.ischool. washington.edu/programs/mlis/, accessed April 24, 2004.
Van House, N., and Sutton, S. A (1996). The Panda Syndrome: an ecology of LIS education. *Journal of Education for Library and Information Science* **37**(2), 131–147.
Van Slyck, A. A. (1995). *Free to All: Carnegie Libraries & American Culture, 1890–1920*. University of Chicago Press, Chicago.
Williamson, C.C. (1923). *Training for Library Service*. New York.
Wilson, T.D. (2004). Review of: Digital library use: social practice in design and evaluation. MIT Press, Cambridge, MA, 2003. *Information Research*, **9**(2), review no. R119 (Available at http://informationr.net/ir/reviews/revs119.html), accessed April 24, 2004.

Index

80/20 rule 111

Abstracting and Indexing (A&I)
  service 109
abstracts 110
academic librarian 161, 158
academic library 5, 9, 11, 14, 27, 105,
  147, 157, 158, 167
Academic Press 99
academic staff 27
Accenture 142
access to materials 5
Active server pages 140
Advancing Learning Communities 92
advertising 101
AIDS 77
ALA 60
All Souls College 4
aluminum 13
Alvarez, Julia 77
Amazon.com 131
ambient light 10
American Academy of Arts and
  Sciences 122
American Library Association 11, 55, 57
American Library Association's
  Committee on Accreditation 180
American Library Association's
  First Congress on Professional
  Education 178
Americans for Libraries Council 13, 55
ANSI/NISO standards 112
apartments 4
architects 1
architect's perspective 1

architectural quality 12
architecture 1, 64
Archives and Records
  Management (ARM) 180
archiving digital content 113
archivists 13, 90
artificial lighting 10, 14
ASIST Best Information
  Science Book Award 186
AskJeeves 160
ASP 140
Association for Library and Information Science Education (ALISE)
  175
Association of Research
  Libraries (ARL) 157
auditorium 6
Australian Business Excellence Award
  (ABEA) 17, 18, 26, 37, 42, 44, 45

Baker Library 5, 6
Barnes & Noble 14
Bates Hall 5
Ben Carson Reading Club 75
benchmarks 25, 42–44, 46
Benton Foundation 142
best practices 26, 42, 49
bibliographic references 112
Bibliothèque National 4
Bibliothèque St Genevieve 4, 5
Big Deal 113
blogs 95
Bodleian Library at Oxford 3
book-reading rooms 3
book stacks 3, 4, 6

Index

bookstores 160
Boston Public Library 5, 56
Boyce 105
branch libraries 59
Breuer, Marcel 7
British Library 114
broadcasters 13, 89
Brown County (WI) Library 57
building digital resources 92
Bureau of Economic Analysis 141
Bureau of Indian Affairs (BIA) 128
Business Excellence Award 13
Business Excellence Framework 22
business organization 13
business principles 27
business results 20, 25

cafes 158
California Digital Library 114
Cambodian Family Journey 67
campus centers 15
Carlitz 145
Carnegie, Andrew 191
Carnegie Corporation 56, 191
Carnegie Library of Pittsburgh 56, 70
carpets 13
carrels 11
cast-iron 4
change agents 18
change management theory 35
change process 35
Charles Stewart Mott Foundation 73
Chemical Online Retrieval
    Experiment (CORE) 99
Chicago Public Library 59
Chicago's Merchandise Mart 11
circulation 154, 157
citizen deliberations 70
city center 3
civic function 12
Civic Library 13, 55, 58, 59, 62–64,
    68, 72, 75, 76, 80
civically oriented service model 13
Cleveland Public Library 76

Cleveland's Senior Success
    Vision Council 77
client 27
client-oriented service organization 13
client relationship 27
client satisfaction 27, 47
Client and Stakeholder Satisfaction 34
client service award 31
Client Service Committee 30
client survey 30
closed stack model 6
coffee shops 158
collaboration 13, 91
collection 3, 6, 9, 19, 20
collection growth 8, 11
collection space requirements 9
collection storage 8
Collegiate Georgian Style 5
comfort 14
commons 64
community information 13
Community Inquiry Labs 62
competitiveness 20
computer 10
computer screens 10
consortium 106
continuing education 60
continuous improvement 21
convenience 14
Cooperative Committee on Library
    Building Plans 6
cost 11
Counter 112
CrossRef 99, 109, 116, 117
cultural communications 12
cultural context 1
cultural heritage institutions 13
cultural institutions 3
cultural value 26
culture of assessment 38, 40
customer 24, 27, 29
customer relationship management 23
customer satisfaction 29
customer service skills 30
Cyber-Infrastructure 187

Cyber-Infrastructure Growth 187
CyberAtlas 156
Cyberliterate 155

data 24
data analysis 33
data and information systems 13
data collection 23
daylight 9, 10
decision making 25
defining quality 21
Delft Technical University 8
democratic social authority 61
Denver 3
Denver Public Library 66
Denver University 14
Department of Energy's Office of Scientific and Technical Information 144
depository library programs 122
Detroit Lakes, MN, Public Library 58
digital formats 171
digital library 192
Digital Library Federation 112
Digital Object Identifier (DOI) 112, 116
digital technology 87
digitization 9
direct sunlight 9
disabilities 125, 141
downtown libraries 3
Drexel University 175, 193
Dublin Core 112

E-Authentication E-Government Initiative 126
E-Citizen 137
e-commerce 123
e-environment 96
E-Government Act of 2002 121, 128, 140
e-government 121, 124, 142
e-mail 159, 164
e-mail messages 161
education 93

education for librarianship 33
Educational Resources Information Center (ERIC) 134
effectiveness 109
electronic-archiving 112
electronic books 19
electronic collections 158
electronic content 104
electronic environment 14
electronic format 19
electronic government 121
electronic information 1
electronic publishing 12, 115
Electronic Records Archive (ERA) 127
employee satisfaction 29
empowerment 36
end-user 158
energy 10
energy use 10
enhancing client 31
entrepreneur 155
environments 11
EnviroMapper 134
ergonomic 10
ERIC 134
Etude de L'Organisation des Bibliothèques 4
Etienne-Louis Boull 4
European modernism 6
excellence 20
expectations 27
expectations of services 20
expectations of the Library 27
eye movement studies 142

facilities 20
FACTER 73
FAST 160
FEDSTATS 130
financial performance 45
FirstGov 129, 131, 137, 142
Flash 129
flexibility 11, 12, 41
flexible space 6
Flint (MI) Public Library 61, 72

Flint Community Networking
 Initiative 61, 72
Flint Timeline Project 73
Florida State University 175
fluorescent 9
fluorescent lights 10
Freedom of Information Act (FOIA)
 128
Freer Gallery of Art of the Smithsonian
 Institution 88
fulltext databases 158
furniture 11

gatekeepers 15
General Accounting Office (GAO) 124
Generation X-ers 159
Geographic Information System (GIS)
 185, 192
Georgia Tech, Virginia Tech 192
Geospatial One-Stop 130
Global Information Infrastructure 187
Google 116, 155
Gordon and Betty Moore
 Foundation 97
GoSaginaw 65
Gould 83
Gov On-line Learning Center 130
government funding 97
Government Performance and Results
 Act 122
Government Printing Office (GPO)
 127, 130, 143
GPO depository library 141
Gunn 145

Halchin 128
halide lamps 9
Harvard Business School 5
Harvard University Library 115
Henri Labrouste 4
Herzog August Bibliotheca 4
higher education 21, 27, 49
Human–Computer Interaction (HCI)
 180
human resource management 26

human resources 23
Human Resources for Information
 Systems Management (HRISM)
 174
hybrids 7

illumination 9
IMLS 84
Impact Factor 115
improvement 22
indicators of success 45
indicators of sustainability 47
indirect lighting 10
industrial relations 23
Informatics 190
Information Access 139
Information Age 1, 13
Information Economics, Management
 and Policy (IEMP) 180
information education 191
information providers 25
Information Renaissance 145
information retrieval 112
Information Science 190
information seeker 14, 147, 151, 156
information services 19
information society 1
Information Systems 190
information technologies 7, 12, 171,
 172, 190
Infotrieve 144
institution of a democratic society 13
Institute of Museum and Library
 Services 13, 83
integrated management system 20
intellectual property 62
interior design 11
intermediated help 153
intermediated service 162
International Standard Book Number
 (ISBN) 112
International Style 6, 7
Internet Protocol (IP) 104
Internet Universe Estimate 156
Internet 2, 14, 86, 95, 154, 161, 167

iProspect 160
iron structure 5

Jacksonville 3
Jacobs 7
Java 129
John Wiley & Sons 14
Johns Hopkins University 15
Johnson County (KS) Public Library 78
journal 8, 108, 115
JSTOR 114

K-12 school system 85
KALIPER 173
Kellogg Foundation 174
Kent State University 186
Key Performance Indicator (KPI) 34
knowledge-driven society 14
Knowledge Economy 86
knowledge management 23
Koninklijke Bibliotheek 114
Koolhaas, Rem 7

Labrouste 4
lamp 9, 10
Laptop 10
Larkin Center 12
Le Comte de Laborde 4
leaders 18
leadership 17, 24, 26, 28, 65, 187
leadership strategies 23
League of Women Voter Forums 79
learning 24, 83, 91
learning environments 13, 66
learning society 84
Leopoldo della Santa 4
LibQual + 27
librarians 13, 90, 97
libraries and learning 83
libraries as business organisations 26
Libraries for the Future 63, 66, 76, 77
Library and Information Science (LIS) 91, 152, 171
Library and Information Science Education 91, 171

library as a physical place 11
library as place 14
library building 12
library-centric perspective 14
library educator 12, 61
library furniture 11
Library of Congress 4
library services 22
library users 14, 162
library's civic role 12
licensed electronic materials 109
licenses 101
life-long learning 12, 69, 87
lighting 9
Likert scale 162
Lipow 151
LIS education 172, 173, 189
LIS literature 161
LIS research 173
LIS scholarship 186
LOCKSS 114
lofts 9
Los Angeles Public Library 67
Louisiana State University Library 88
Lucker, Jay K. 8

maintenance 13
MAJOUR (Modular Application for Journals) 99
Many Voices, Many Lives 77
Marchionini 142
market forces 96
marketing 27
Maxine Hong Kingston 77
MBA 161
MBA students 148, 152
McCabe, Ronald 61
McClure 123
McCook, de la Peña 61
McKim, Mead, and White 5
media 9, 14
Mellon funding 114
mental models 152
Metadata Harvesting 116
metal halide fixtures 10

Mies Van der Rohe 7
Millenials 159
MIT 192
modernist design 7
modular library 6
modular plan 7
monograph publication 8
morale 22
Mu 142
Multitaskers 155
museum curators 13
museum professionals 90
museum–library collaborations 92
museums 88
Myers–Briggs Type Indicator®
 (MBTI) 31

NASA 136
Nashville 3
National Atlas Online 133
National Council of Jewish Women 71
National Gallery of Art 88
National Institute for Standards and Technology (NIST) 129
National Issues Forums 66, 79
National Leadership Grants 83, 92
National Library of the Netherlands 114
National Performance Review 123
National Women's Health Information Center of the Department of Health and Human Services 131
natural daylight 9
natural light 14
Neal-Schuman 182
NEOCON 11
NERL 97
NESLI 97
New Haven (CT) Free Public Library 15, 74
New Haven's Non-Profit Resource Center (NPRC) 75
New York University Libraries 15
non-commercial setting 27

North Coast Seniors Connection 77
North light 9
not-for-profit organisations 17, 26

OCLC Environmental Scan 86, 153, 159
Odlyzko 105
Office of Electronic Government (OEG) 145
Office of Management and Budget (OMB) 121
offices for librarians 3
offices for staff 4
Ohio Library Foundation's Drew Carey Young Adult Service Program 76
OhioLINK 97, 99, 106, 109, 113
online library 160
Open Access 97, 117
Open Archives Initiative Protocol 116
Open Society Institute 100
OpenURL 112
organizational culture 38
organisational growth 22
organisational learning 18
organisational outcomes 23
organisational performance 25
Otis Elevator factory 12
outcomes 20, 25
output measures 25

Paris 4
Pasadena (CA) Public Library 57
partnerships 13
patron seating 6
Paul Allen 97
pay-per-view 108
People's University 76
performance 22
Performance Indicator Framework 28
performance indicators 25, 27, 50
performance management 23, 25
performance management system 19

Personal Data Assistants (PDAs) 110, 154
personal information economy 153
Pew Internet & American Life Project 153, 156
physical places 2, 12
planning 24
processes 24
planning processes 23
Portable Document Format (PDF) 99
portals 129
post-war design 7
preservation 14
Pricing Electronic Access to Knowledge (PEAK) 99
pricing models 12
pricing terms 111
Principles of Business Excellence 23
process improvement 13, 22, 46
professional education 12
professional societies 97
Project for Public Spaces (PPS) 58
Providence Public Library 67
public affairs programs 66
public and academic libraries 1
public areas 64
public broadcasting 87
public dialogue 13
public good 49, 64
public good organisations 26
public library 9, 11, 13, 14, 56, 65, 90, 152, 160
Public Library Association 59
Public Library of Science (PLoS) 98
Public Library of Science and BioMedCentral 117
public meetings 6
public memory 13, 66
public programs 3
public reading spaces 4
public space 1, 13
publisher's perspective 13
publishers 95, 97
PubScience 144
punctuated equilibrium 83

quality 21
quality assurance 17, 21
quality audit 21
quality award 12
quality concepts 27
quality education 69
quality framework 22
quality improvement 13
quality journey 13, 18, 20, 22, 30, 33, 42
quality management 17, 21, 35
quality of life 12
quality process 33
quality program 18, 21
Quality Service Excellence 21
Queens Borough (NY) Public Library 66
Questia.com 160

Radcliffe Library 4
Raphael's School of Athens 4
readers 4, 5, 155
readership 156
reading room 3–6, 12
reading spaces 7
Readmobile 74
Recruitment One-Shop 130
Red Sage 99
reference desk 159
reference queries 157
reference requests 158
reference services 158
Rehabilitation Act 138
Relyea 123
renewable materials 13
renovations 12, 13
Research and Demonstration 92
research libraries 9
researchers 12
return on investment 26
Robert A.M. Stern Architects 1, 3, 6
Rodski Customer Survey 27, 28
role of leadership 18
role of the library in publishing 14
Ronald McCabe 61
rooms for readers 4

Royal Library 4
Rubin 186

Saginaw (MI) Public Library 65
Saginaw Community Connection 65
Salt Lake City 3
Salt Lake City (UT) Public Library 64
San Antonio 3
satisfaction 86
satisfied clients 39
scholarly communication process 97
scholarly communications 12
scholarly publishers 95
Scholarly Publishing and Academic Resources Coalition (SPARC) 98
Scholar's Portal 98
scientific communication 115
seamless 87
seamlessness 86
search engine 155
Seattle 3
Seattle Public Library 7
self-help 153, 156
self-service 86
self-sufficiency 86
serial 19
Serial Item and Contribution Identifier 112
service 24
service excellence 28
service orientation 20
service satisfaction 29
SERVQUAL 141
Shades of LA 67
shelve books 13
Simmons College 14
skylights 9
social context 13
social evolution 3
Soros, George 97
spaces 1, 12, 19
special libraries 152
special and academic libraries 159
St Genevieve 4

stack core 5, 6
stack lights 10
stacks 10
staff 20
staff development 19, 23
staff resistance 22
stakeholders 24, 25, 27, 49, 52, 95
Standard Generalized Markup Language (SGML) 99
Starbucks 14
steel 13
STM 100
STM journal publishing 96
storehouses of knowledge 14
strategic plans 19, 24, 26, 28, 40
students as customers 27
Study Circles 79
study-space 19
subscriptions 101
Suffolk University 14
sustainable design 12
SWOT analysis 40
Syracuse 193
systems 24

tables of contents 110
task light 9
Tacoma (WA) Public Library 57
Taylor 152, 173
Taylor model 156
Taylor, Robert 172
team building 38
teamwork 38, 155
technical invention 3
technological change 20, 26
technology 11, 22, 24, 155, 167, 172
techno-positivist school 7
Texas State Library and Archives 88
The Commons 57, 59
The Library of the Future 69
The University Licensing Program (TULIP) 99
Thomas, Piri 77
Three Rivers Free-Net 71
total quality control 17

total quality management 17, 21
total quality service 21
TQM 21, 22, 27, 38
transformational changes 18
trustees 12

UCLA 172
UNESCO public library manifesto 56
universities 5, 21, 25
university administrators 27
university libraries 104
University of California, Berkeley 178
University of California, Los Angeles 178
University of Cincinnati 99
University of Colorado 161
University of Illinois 171
University of Kentucky 154
University of Kentucky School of Library and Information Science 186
University of Leyden 3
University of Maryland at College Park 14
University of Michigan 175, 193
University of Michigan School of Information Science 72, 187, 193
University of North Carolina—Chapel Hill 188
University of Pittsburgh 193
University of Texas, Austin 178
University of Virginia 4
University of Washington 178
University of Wisconsin, Milwaukee 190

University of Wollongong 18
University of Wollongong Library 13, 17
US Green Building Council, 13
usage data 14, 109, 110
usage-based pricing 113
use of energy 14
user-centric view 14
user education 165

value 26, 109
Vancouver Library 58
Venturi 7
Virginia Beach Public Library 69
virtual patron 103
virtual reference desk 163, 164

Walter Gropius 6
warehouse 2, 12
Web 14, 96, 108, 116
WebGov 129
Wedge, Carole 58
Williamson 191
Williamson Report 176, 191
World Affairs Council of Pittsburgh 71
World AIDS Day 77
World Wide Web 166

Yread 76
Yonkers Public Library 12

Zaha Hadid 7
Zweizig 90, 151